CLIMATE OF THE SOUL

OTHER BOOKS BY ANDREW D. MAYES

spiritualityadviser.com

Celebrating the Christian Centuries (1999)

Spirituality of Struggle: Pathways to Growth (2002)

Spirituality in Ministerial Formation (2009)

Holy Land? Challenging Questions from the Biblical Landscape (2011)

Beyond the Edge: Spiritual Transitions for Adventurous Souls (2013)

Another Christ: Re-envisioning Ministry (2014)

Learning the Language of the Soul (2016)

Journey to the Centre of the Soul (2017)

Sensing the Divine (2019)

Gateways to the Divine: Transformative Pathways of Prayer from the Holy City of Jerusalem (2020)

Diving for Pearls: Exploring the Depths of Prayer with Isaac the Syrian (2021)

Voices from the Mountains: Forgotten Wisdom for a Hurting World from the Biblical Peaks (2021)

CLIMATE OF THE SOUL

Ecological Spirituality for Anxious Times

ANDREW D. MAYES

RESOURCE *Publications* • Eugene, Oregon

CLIMATE OF THE SOUL
Ecological Spirituality for Anxious Times

Copyright © 2022 Andrew D. Mayes. All rights reserved. Except for brief quotations in critical publications or reviews, no part of this book may be reproduced in any manner without prior written permission from the publisher. Write: Permissions, Wipf and Stock Publishers, 199 W. 8th Ave., Suite 3, Eugene, OR 97401.

Resource Publications
An Imprint of Wipf and Stock Publishers
199 W. 8th Ave., Suite 3
Eugene, OR 97401

www.wipfandstock.com

PAPERBACK ISBN: 978-1-6667-3713-4
HARDCOVER ISBN: 978-1-6667-9624-7
EBOOK ISBN: 978-1-6667-9625-4

03/09/22

Unless otherwise acknowledged, Scripture quotations are from New Revised Standard Version Bible, Copyright © 1989, 1995 National Council of the Churches of Christ in the United States of America. Used by permission. All rights reserved worldwide.

Revised Standard Version (RSV) Copyright © 1946, 1952, and 1971 National Council of the Churches of Christ in the United States of America. Used by permission. All rights reserved worldwide.

Jerusalem Bible (JB) Copyright © 1966 Darton, Longman & Todd.

New Living Version (NLV) Copyright © 1969, 2003 Barbour Publishing, Inc.

The Voice Bible (VOICE) Copyright © 2012 Thomas Nelson, Inc.

"Wind, wind , blow on me" Jane & Betsy Clowe ©1974,1975 Celebration/Kingsway Thankyou Music.

"Great is Thy faithfulness" Thomas Obediah Chisholm (1866–1960) © 1923. Ren. 1951 Hope Publishing Co., Carol Stream, IL 60188 www.hopepublishing.com

Contents

Introduction		vii
1	Mystic Metaphors: Reading the Soul	1
2	Interpreting the Universe: Pilgrims, Prophets and Poets	12
3	Facing the Elements: Jesus' Experience and Message	24
4	Welcoming the Unpredictable: Seasons of the Spirit	38
5	Embracing Transition: Winds of Change	47
6	Braving Exposure: The Summons of the Sun	59
7	Entering the Cloud: Mystery and Presence	76
8	Thundering in the Soul: Protest and Silence	88
9	Replenishing the Spirit: Rain, Flood and Drought	103
10	Discovering the Cosmic Christ: Filling all Things	115
Bibliography		131

Introduction

The voice of the Lord is upon the waters. The God of shining-greatness thunders. The Lord is over many waters. The voice of the Lord sends out lightning. The voice of the Lord shakes the desert. The Lord sits as King over the flood.
(Ps 29: 3-8, 10, NLV)

May all the children-to-come fear You as long as the sun and the moon last. May He come down like rain upon the cut grass, like rain that waters the earth. In His days may all go well with those who are right and good. May there be peace until the moon is no more.
(Ps 72: 5-7, NLV)

This book bears us on eagles' wings into the vault of the heavens and plunges us into the hidden depths of the soul. At the same time as the climate crisis alerts us to the state of the planet, so we look into our soul. Little attention has been given to how the Bible and spiritual writers through the centuries use arresting meteorological imagery to describe both the discovery of the Divine and the condition of the mortal. This book invites us to explore a rich and diverse vocabulary, archetypal, universal and primal, which enables us to describe the movements of the soul. These images and metaphors help us give expression to what is going on in our spiritual lives, as we learn the skill of reading the climate of our soul.

We'll face the challenge to respond to change and to welcome transformation throughout the course of the spiritual journey. This is a summons to risk a radical exposure to God. It renews the call to spiritual adventuring.

We begin each theme by examining biblical material. We explore each theme by reconnecting to classic spiritual writers in the Christian tradition. We allow ourselves to be heartened, challenged and energized by our discoveries.

Above all, we seek to deepen more reverential attitudes to all of created life, including the elements, an appreciation of the sacredness of everything including the physical climate and its movements. We seek to develop our awareness of the universe as a revelation of God, primary "sacred scripture," as we discover a sense of the cosmic Christ embracing the heavens and the earth. Attentiveness to the ecology of the soul will lead us to a sharper perception of the environmental issues facing our planet. Physicality points to spirituality—and vice versa. This book is an invitation to a different way of seeing—celebrating sacramental approach to the universe, to the elements. This book aims to be

- A catalyst and stimulus to adventurous spirituality
- A tool to help us read and make sense of our faith journey and our turbulent world
- A resource which brings us back to the wellsprings of spirituality in scripture and spiritual writings
- A key, unlocking a fresh interpretation of the transitions we face
- A summons to living with greater attentiveness and understanding to the changing climate

USING THIS BOOK

The book is designed to be used by both individuals and groups. Questions at the end of each chapter are provided to stimulate personal reflection and group discussion. Three readerships are in mind. First, it is for those longing for movement and progress in their spiritual lives. Second, it is for those who support others on their spiritual journey: those who serve as spiritual directors, soul-friends or accompaniers. Third, it is for seekers, for those wanting to discover for themselves the astonishing riches of classic spiritual writers. The book will open the user to a wide variety of spiritual resources that will inspire the spiritual journey. It can be used either alone or in house-groups. It is recommended that both individuals and course participants keep a journal or note book, in which to note and reflect on the transitions taking place in themselves as they undertake this life-changing journey.

"The world is at one minute to midnight," having run down the clock on waiting to combat climate change, UK Prime Minister Boris Johnson declared at the opening in November 2021 of the COP26 Climate Change Conference.[1] We live in anxious times. UN Secretary General Antonio Guterres concluded the conference with the words: "Our fragile planet is hanging by a thread. We are still knocking on the door of climate catastrophe."[2] We have finally woken up to the devastating effects of climate change. We know what aggravates global warming, melting of icecaps and rising sea levels. Scientists analyze the symptoms and identify possible causes. We are becoming aware of complex contributory factors and alert to the changes that are taking place in our very lifetimes. The very air we breathe into our lungs, especially in our cities, bears poisonous particles from car exhausts, while factories pump pollutants high into the air that may fall as acid rain. With our CFCs, we deplete the ozone layer that protects us from harmful ultraviolet rays of the sun. The climate is demanding our attention as a species, calling us to interpret the weather...

In his 1892 novel *The American Claimant* Mark Twain delights to say: "No weather will be found in this book. This is an attempt to pull a book through without weather. It being the first attempt of the kind in fictitious literature, it may prove a failure, but it seemed worth the while of some dare-devil person to try it, and the author was in just the mood." Twain goes on to admit: "Of course weather is necessary to a narrative of human experience." Today most Americans follow weather events closely and with concern, fearing the devastating effects of tornadoes, hurricanes, droughts and floods. A record number of Americans are anxious about global warming: numbers are rising significantly.

On a lighter note, on the other side of the Pond, the BBC has promoted its weather-watchers with the slogan: "the nation's favorite conversation." The Daily Telegraph reported, prior to the pandemic: "The weather is still Britain's favorite topic of conversation with three quarters of us discussing it more than anything else, according to research."[3] In the eighteenth century the poet and writer Samuel Johnson (1709–84) observed: "When two Englishmen meet, their first talk is of the weather; they are in haste to tell each other what each must already know, that it is hot or cold, bright or cloudy." Today, on getting up the very first question many people ask is: "what is

1. Rowlatt, "COP26."
2. Guterres, "COP26 Statement."
3. Telegraph, "Weather."

the weather doing today?" The weather affects our moods and behaviors. It can have productive or debilitating effects on our health. It determines our choice of clothing. It shapes agriculture and food production. It influences our planning and organizing—in the UK we often need to have a contingency plan for outside summer events, should the weather disrupt our hoped-for arrangements. More seriously, in many parts of the world adverse weather can bring flooding or drought, famine or plenty. The environmental conditions hovering above the surface of our planet have their impact on almost everything that humans need to do.

English literature brims with the imagery and symbolism of the weather. King Lear, Shakespeare tells us, "was minded like the weather"—his turbulent behavior reflecting nearby storms. The very title of Emily Bronte's novel *Wuthering Heights* suggests what is to come: the word "wuthering" referring to a wind so strong that it makes a roaring sound—such threatening weather creates a sense of foreboding and expectation of stormy characters and tumultuous relationships. In Dicken's *Bleak House* it rains for the first twelve chapters, before pausing and raining again! In *The Lion, the Witch, and the Wardrobe* Narnia was taken over by the evil Snow Queen, blanketing the land in a never-ending winter (with no Christmas!) During winter, nothing grows, and all seems dead or sleeping. But when Aslan returns, the snow begins to melt and spring arrives: hope for the world has returned. We recall, too, movies that utilize weather imagery evocatively and atmospherically, in everything from *The Great Gatsby* to *Game of Thrones*. But while we have knowledge of these usages of weather imagery, what do we know of how the Bible employs this powerful and evocative symbolism?

In the Bible vivid and arresting images of weather furnish us with a rich vocabulary we can use to describe both the divine workings and our own spiritual life. Physicality points to spirituality, and meteorology points to cosmology—ways of reading the world, and ways of reading our own soul. Hildegard of Bingen sings: "The sum total of heaven and earth . . . becomes a temple and altar for the service of God."[4] This book is an invitation to rediscover a spiritual universe and a symbolic universe. The biblical cosmology summons us to an ancient, yet ever fresh, symbolic perception of the world that alerts us in new ways to our environment now gravely under threat.

In earlier books, I explored the spiritual life through the imagery of the visible landscape. In *Holy Land? Challenging Questions from the Biblical Landscape* I led the explorer across the terrain of the Holy Land, which I had got to know well when working as course director at St George's

---

4. Uhlein, *Meditations with Hildegard*, 107.

College, Jerusalem, and through my ongoing ministry in regularly leading pilgrimages. There we explored the mountains, rivers, gardens, deserts and ocean and allowed the physicality of the land to throw at us vital questions and raise thorny issues in spirituality. In *Beyond the Edge: Spiritual Transitions for Adventurous Souls* we followed Jesus into liminal spaces across the Land—venturing to the coastlands, entering "no-go areas," wading across the Jordan—unpacking the theme of crossing boundaries in order to experience at once a radical letting-go and a startling rediscovery of the spiritual life. *Voices from the Mountains: Forgotten Wisdom for a Hurting World* took us up to the heights of the biblical peaks, while in *Journey to the Centre of the Soul: a Handbook for Explorers* we left the surface terrain of the Holy Land and ventured underground, exploring the spiritual life through the extended metaphor of subterranean and cave spirituality. Now it is time to look heavenwards, and discover clues to the soul in the environment and atmosphere. This present book builds on my exploration of metaphor in spirituality in *Learning the Language of the Soul* (2016). We seek to discover the approach exemplified by English priest and poet Thomas Traherne (1637–74) when he wrote: "Of hills and mountains, rain and hail, and snow, clouds, meteors etc. how apparently the Wisdom, and Goodness, and Power of God do shine in these."[5]

THE VARIED AND UNPREDICTABLE CLIMATE OF THE BIBLICAL LANDS

Because of its geographical setting and landforms the climate of the Holy Land is not predictable. Within a relatively short distance, there are several contrasting climatic zones. Fringing the Mediterranean lies the flat coastal strip. As one proceeds inland, one is confronted by the rugged central highlands of Judea and Samaria, Jerusalem itself set in the midst—"as the mountains are round about Jerusalem so the Lord is round about his people" (Ps 125:2). To the east of the watershed of the Mount of Olives the steppe quickly passes to the bleak, rocky canyons of the Judean Desert. The land drops away from a height of 2700 feet above sea level to 1250 feet below, to the deepest place on earth, the great rift valley, where the Jordan flows into the Dead Sea. The desert is raw, wild, untamed terrain, eroded by elements of wind and sun and water, splitting rocks and crumbling cliffs, symbolizing the brokenness of humanity; it is an open, exposed place, bespeaking of the vulnerability of the soul.

5. Traherne, *Kingdom*, 400.

The Jordan begins in the far north as the meltwaters of the snow-capped Mount Hermon, which rises to almost six thousand feet. From Dan to Beersheba, the ancient descriptors and perimeters of the Holy Land, there could not be a greater contrast—Dan being set in the mountains of the Upper Galilee, bordering Syria, while Beersheba in the south marks the edge of the bleak Negev desert covering 55 percent of the country's area, which extends 150 miles until it reaches the shores of the Red Sea.

The early Hebrew setters noticed a significant contrast with the predictable climate of the Egypt they were fleeing in the Exodus. That climate had its regular routines due to the seasonal flooding of Nile, as it spread out its rich silt. But Palestine was to be quite different:

> For the land that you are about to enter to occupy is not like the land of Egypt, from which you have come, where you sow your seed and irrigate by foot like a vegetable garden. But the land that you are crossing over to occupy is a land of hills and valleys, watered by rain from the sky, a land that the Lord your God looks after. The eyes of the Lord your God are always on it, from the beginning of the year to the end of the year. (Deut 11:10–12)

Victorian explorer of the Holy Land George Adam Smith observed: "In the Palestine year there is no inevitableness. Fertility does not spring from a source which is within control of man's spade . . . a purely mechanical conception of nature as something inevitable, whose processes are more or less under man's control, is impossible . . . the climate of Egypt does not suggest a personal Providence, but the climate of Palestine does so."[6] A recent scholar echoes this thought. In his important study *The Natural History of the Bible* Daniel Hillel affirms: "The basic reason for the Israelites' troubles lay in the environment where, by historical coincidence, they staked out their life as a nation. It is a land of unstable climate, at the edge of the desert, with some years or succession of years blessedly rainy and others accursedly dry . . . The land became a sort of moral seismograph, an indicator of the nation's collective behavior. Its manifestations were to be watched at all times for telltale signs of the return of the desert."[7]

This suggests an important theme as we explore the imagery of climate: like the very climate of the Holy Land itself, the spiritual life is not predictable, but capable of many different types of development. Nothing is predetermined or fixed in advance except the constant invitation to greater Christlikeness. Like the weather itself, we should be changeable in the sense of being able to respond in different ways to God and human need. The

---

6. Smith, *Historical Geography*, 72, 73.

7. Hillel, *Natural History of the Bible*, 214, 215. See also, Simkins, *Creator and Creation*.

practice of prayer changes the atmosphere of our heart and we experience shifts in the weather patterns of the soul. We need rule nothing out. We need not become trapped in routines and regularities if they are becoming unfruitful. God is always summoning us forwards, into an adventurous unpredictable life in the Spirit. And further: his providence will not fail us. There is, as it were, a reciprocal relationship between heaven and earth:

> "On that day I will answer your prayers," declares the Lord.
> "I will speak to the sky,
> it will speak to the earth
> and the earth will produce grain, new wine, and olive oil.
> You will produce many crops, Jezreel." (Hos 2.21, 22)

There is a double theme running through the Scriptures:

CLIMATE SUGGESTS IMAGES OF GOD AND REPRESENTS DIVINE ACTION

Isaiah, for example, delights in imagery from the weather to depict the work of God:

> Shower, O heavens, from above,
> and let the skies rain down righteousness;
> let the earth open, that salvation may spring up,
> and let it cause righteousness to sprout up also;
> I the Lord have created it. (45:8)
> For as the rain and the snow come down from heaven,
> and do not return there until they have watered the earth,
> making it bring forth and sprout,
> giving seed to the sower and bread to the eater,
> so shall my word be that goes out from my mouth;
> it shall not return to me empty,
> but it shall accomplish that which I purpose,
> and succeed in the thing for which I sent it. (55:10, 11)
> For thus the Lord said to me:
> I will quietly look from my dwelling
> like clear heat in sunshine,
> like a cloud of dew in the heat of harvest. (18:4)

CLIMATE SUGGESTS SELF-IMAGES AND UNDERSTANDINGS OF VOCATION

It is also a mirror of the soul. The weather interprets our soul, and reflects our state of mind. The external world provides us with images with which we express the interior world. For example Jonah's story describes a violent storm: "the Lord hurled a great wind upon the sea, and such a mighty storm came upon the sea that the ship threatened to break up" (Jonah 1:4). The author of 2 Kings uses the same Hebrew world for storm to describe the anxiety of the perturbed king: "Then the heart of the king of Aram was stormy because of this matter" (2 Kgs 6:11, Lexham English Bible). In a psalm of lamentation, describing horror at a friend's treachery and betrayal, the storm becomes a metaphor for fear:

> I am distraught by the noise of the enemy,
>   because of the clamor of the wicked.
> For they bring trouble upon me,
>   and in anger they cherish enmity against me.
>   I would hurry to find a shelter for myself
>   from the raging wind and tempest. (Ps 55:3,8)

THE IMPACT OF THE CLIMATE ON SPIRITUALITY

Little attention or research has been given to the changing climate which forms the background to the remarkable flowering of mysticism in England and across Europe in the fourteenth century. Time and again, the spiritual writers of this period allude to the weather. Bitterly cold winters and drenching rains in the early 14th century announced the end of the Medieval Warm Period, heralding the dawn of what has been called "the Little Ice Age." The weather was exceptionally variable between the 1280s and the 1420s, indicating significant shifts in the climatic system of the northern hemisphere. The Baltic Sea froze over twice in 1303 and 1306-7 and in 1309/10 the Thames iced over. Polar and alpine glaciers were advancing.[8] Incessant rains in the 1310s caused agricultural failures across Europe and famine became widespread.

Wrapped in a thick woolen cloak and shivering in her cell in Norwich as biting winds from the North Sea swept across the East Anglian flatlands, Julian wrote her consoling words: "He is our clothing who for love wraps and encloses us." She describes her vision of Jesus' flow of blood in his passion in

---

8. Tuchman, *Distant Mirror*. See also, Brian Fagan, *Little Ice Age*.

these terms: "The copiousness resembles the drops of water which fall from the eaves of a house after a great shower of rain, falling so thick that no human ingenuity can count them . . . .This vision was living and vivid."[9] Later, she relates: "For at the time when our blessed savior died upon the Cross, there was a dry, bitter wind, I saw."[10]

In bitter, plunging temperatures the English mystic Richard Rolle (1300–49) chooses as his main metaphor and as the title of his key work *The Fire of Love*. It opens with the arresting words: "I cannot tell you how surprised I was the first time I felt my heart begin to warm. It was a real warmth too not imaginary, and it felt as if it were actually on fire . . . But once I realized that it came entirely from within, that this fire of love had no cause, material or sinful, but was the gift of my Maker, I was absolutely delighted . . . " But he says that such a flame comes and goes:

> It will reappear in time, though until it does I am going to be spiritually frozen . . . When I am awake I can try to warm my soul up, though it is numb with cold . . . I catch myself growing cold: cold until once again I put away all things external, and make a real effort to stand in my Savior's presence: only then do I abide in this inner warmth.[11]

Rolle extols, with unbridled enthusiasm, the affective dimensions of Christian spirituality: his emotional states seem to reflect in some way the external environment. In less optimistic mode, Walter Hilton (1340–96) alludes to England's bleak weather: "This light of false knowledge shown by the devil to a soul in darkness is always seen between two black rainclouds. The higher cloud is presumption and self-conceit, while the lower is oppression and depreciation of our neighbor . . . errors and heresies pour from them like rain from black clouds."[12]

Across the continent of Europe, spiritual writers reflect the bitter cold of the environment. The Italian seer Catherine of Siena (1347–80) writes of the imagery of fire in a very powerful way:

> You know the only thing that can bind a person is a bond; the only way to become one with the fire is to throw oneself into it that not a bit of oneself remains outside it . . . Once we are in its embrace, the fire of divine charity does to our soul what physical

---

9. Colledge & Walsh, *Showings*, 188.

10. Colledge & Walsh, *Showings*, 206.

11. Wolters, tr., *Fire of Love*, 45,46 . See also Allen, tr., *Richard Rolle*; Martin, *English Spirituality* ; Reveney, *Language, Self and Love*. Rolle has been called "the father of English literature."

12. Sherley-Price, *Walter Hilton*, 17.

> fire does; it warms us, enlightens us, changes us into itself. Oh gentle and fascinating fire! You warm and you can drive out all the cold of vice and sin and self-centeredness! This heat so warms and enkindles the dry wood of our will that it bursts into flame and swells in tender loving desires, loving what God loves and hating what God hates. And I tell you, once we see ourselves so boundlessly loved, and see how the slain lamb has given himself on the wood of the cross, the fire floods us with light, leaving no room for darkness. So enlightened by that venerable fire, our understanding expands and opens wide. For the light from the fire lets us see that everything (except sin and vice) comes from God . . . Once your understanding has received the light from the fire as I've described, the fire transforms you into itself and you become one with the fire . . . How truly then we can say that he is a fire who warms and enlightens and transforms us into himself![13]

Meister Eckhart (1260–1328) also remains optimistic as he places his trust in God:

> The poor one said, "You wish me a good day." I have never had a bad day. For if I was hungry I praise God; if it freezes, hails, snows, rains, if the weather is fair or foul, still I praise God; am I wretched and despised, I praise God, and so I have never had an evil day. You wish that God would send me luck. But I never had ill luck, for I know how to live with God, and I know that what God does is best; and what God gives me or ordains for me, be it good or ill, I take it cheerfully from God as the best that can be, and so I never had ill luck. You wished that God would make me happy. I was never unhappy; for my only desire is to live in God's will, and I have so entirely yielded my will to God's, that what God wills, I will.

FOUR INVITATIONS OF THIS BOOK

1 Invitation to Radical Exposure

For millennia humans have lived in a state of exposure—lashed by wind and rain. Today we live in insulated homes, and travel unexposed, not on horse or donkey, but in cars with heaters and impenetrable windows. In our time, we have become disconnected, removed, detached from the vagaries

---

13. Letter 51, to Apostolic Nuncio to Tuscany.

of weather and protected from its impact. Today we resolve "we won't let the weather affect us," trying to reign supreme even over the elements—a symbol of our desire to maintain control over our own lives? We are not called to be "fair weather Christians" but to embrace the life of faith, which is a summons to risky living. As we follow this theme in scripture and writings, we find ourselves beckoned to rediscover a certain vulnerability and radical exposure to God. Gerard Manley Hopkins asks:

> What would the world be, once bereft
> Of wet and of wildness? Let them be left,
> O let them be left, wildness and wet;
> Long live the weeds and the wilderness yet. (*Inversnaid*, 1918)

This book summons you to spiritual nakedness and utter exposure to God, to quit the safe confines of domesticated living. We are so used to protecting ourselves from the raw elements: our clothing, housing, heating systems insulate us from the raw impact of the weather and distance us from the elements which our ancestors encountered in all their wildness. We retreat to artificial air-conditioned thermostatically-controlled environments. This symbolizes the way we erect self-protective barriers against the Divine—to keep God at arms' length—we want God close, but not *too* close! A domesticated consciousness represents a narrowing of the spirit, and seeks to place a limit on the Divine: we wear our spiritual armor without realizing it. This book calls us to lower our self-protective veiling and come face-to-face with the utter mystery and wonder of God. We are invited to an expansiveness of spirit—and maybe we can face our own wildness too!

This book also invites you to an unpredictable life—as uncertain as the weather. Sure, certain patterns can be discerned and identified, but there is no long-term forecast in the spiritual life. We are invited to step out into the gale of the Spirit, to be redirected and re-energized by the very breath of God. We discover the experience of being inundated and engulfed by the wild waters of the Spirit. We lose ourselves in the dense cloud of God's mysterious presence. This book may unnerve, shake up, unsettle, disturb, provoke. We need to be ready to change, to move forward with the elemental energies of God. It may also bring us fresh perspective on our spiritual lives, and help us clarify and name what is going on in our lives. It opens up for us interpretative frameworks by which we can decipher and make sense of the spiritual adventure.

2 Invitation to Attentiveness

Thomas Merton (1915–68) calls us to a high state of wakefulness:

> It might be a good thing to open our eyes and see.
> It is essential to experience all the times and moods of one good place.
> It is God's love that warms me in the sun and God's love that sends the cold rain. It is God's love that feeds me in the bread I eat and God's love that feeds me also by hunger and fasting . . . It is God who breathes on me with light winds off the river and in the breezes out of the wood.
> As we go about the world everything we meet and everything we see and hear and touch . . . plants in us . . . something of heaven. It is good and praiseworthy to look at some created thing and feel and appreciate its reality. Just to let the reality of what is real sink into you . . . for through real things we can reach Him who is infinitely real . . .
> The sun on the grass was beautiful. Even the ground seemed alive . . . [14]

In *The Sign of Jonas* Merton celebrates the formative influence of the weather on the soul:

> How necessary it is for monks to work in the fields, in the rain, in the sun, in the clay, in the wind: these are our spiritual directors and our novice-masters. They form our contemplation. They instill us with virtue.[15]

He calls us to a greater attentiveness and awareness of mind and soul:

> Our mentioning of the weather—our perfunctory observations on what kind of day it is, are perhaps not idle. Perhaps we have a deep and legitimate need to know in our entire being what the day is like, to *see it* and *feel it*, to know how the sky is grey, paler in the south, with patches of blue in the southwest, with snow on the ground, the thermometer at 18, and cold wind making your ears ache. I have a real need to know these things because I myself am part of the weather and part of the climate and part of the place, and a day in which I have not shared truly in all this is no day at all. It is certainly part of my life of prayer . . .
> . . . .look at your own life, and your part in the universe as infinitely rich, full of inexhaustible interest, opening out into

---

14. Merton, quoted in De Waal, *Seven Day Journey*, 20.
15. *Sign of Jonas* quoted in Merton, *Trees*.

infinite possibilities for study and contemplation and interest and praise. Beyond all and in all is God . . . .

Forest and field, sun and wind and sky, earth and water, all speak the same silent language, reminding the monk that he is here to develop like the things that grow all around him.[16]

As we engage with the message of the elements, we start to make greater sense of our vocation. We develop an awareness that becomes transformative—as Pope Francis puts it:

> The universe unfolds in God, who fills it completely. Hence, there is a mystical meaning to be found in a leaf, in a mountain trail, in a dewdrop, in a poor person's face. The ideal is not only to pass from the exterior to the interior to discover the action of God in the soul, but also to discover God in all things.[17]

3 Invitation to Double Vision

Poet and visionary William Blake invites us to employ a kind of double vision:

> For double the vision my eyes do see,
>   And a double vision is always with me:
>   With my inward eye 'tis an old man grey;
>   With my outward a thistle across my way.

And he goes on:

> May God us keep
>   From single vision and Newton's sleep![18]

He invites us to move beyond Newton's mechanistic view of the universe to greater insight. In *The Last Judgment* he faces the question: "What, when the sun rises, do you not see a round disc of fire, somewhat like a guinea?" He responds: "Oh! no! no! I see an innumerable company of the heavenly host, crying, 'Holy, holy, holy is the Lord God Almighty!' I question not my corporeal eye, any more than I would question a window concerning a sight. I look through it, and not with it." George Herbert had written in similar vein:

---

16. Extracts from "Turning Toward the World," "A Search for Solitude," "Waters of Siloe" in Merton, *Trees*.

17. Francis, *Laudato Si'*, para 233.

18. "Letter to Thomas Butts, 22 November 1802" in Erdman, *Complete Poetry*, 720.

> A man that looks on glass,
> on it may stay his eye;
> or if he pleaseth, through it pass,
> and then the heaven espy.

Blake invites us to discover a symbolic universe, opening up doors of perception. Frye says of Blake's verse: "the conscious subject is not really perceiving until it recognizes itself as part of what it perceives. The whole world is humanized when such a perception takes place."[19] Meanwhile, Herbert wakes us up to a deeply sacramental view of life—a new way of seeing reality, glimpsing the presence of God in all things, which can be transformative.

In the architecture of the earliest Church, the first thing the first Christians saw as they emerged, dripping, from the baptismal pool, to begin their new life, was a depiction of heaven, with its stars and firmament.[20]

The letter to the Colossians also issues a clear summons to us to direct our gaze heavenwards:

> So if you have been raised with Christ, seek the things that are above, where Christ is, seated at the right hand of God. Set your minds on things that are above, not on things that are on earth, for you have died, and your life is hidden with Christ in God. When Christ who is your life is revealed, then you also will be revealed with him in glory. (Col. 3:1–4)

Christian spirituality became infected with divisive, dualistic thinking since the early centuries embraced Platonic thought. Plato himself wrote in the *Republic* that attentiveness to the world of the senses was "looking in the wrong direction."[21] Disastrous polarities crept into Christian thinking, undermining the idea of God's incarnation. Things were pitched against one another: heaven was opposed to earth, the body to the spirit. Politics and prayer were to be kept separate. Sacred and secular were delineated with barriers, as if they were two separate realms, holy and unholy. The church and the world are set against each other. In Christian thought and practice, the spiritual world was given primary attention as being more important than the physical, and the natural world viewed as being only of secondary or temporary significance. In spirituality, such dualistic thinking has created unnecessary distances and opened up uncalled-for chasms. Robert

---

19. Frye, *Double Vision*, 23.

20. As evidenced at the earliest surviving Christian church of Dura-Europos, Syria described in White, *Lost Knowledge*, 70.

21. Plato, "Republic," quoted in Louth, *Origins*, 6.

McAfee Brown subtitled his book *Spirituality and Liberation* with the words: "overcoming the great fallacy."[22] He identifies this as a persistent dualism that separates and opposes faith and ethics, the holy and the profane, the otherworldly and this-worldly, eroding the central Christian belief in the Word made flesh. Incarnational spirituality celebrates not only God within, but God in our very midst, in the dirt and in the gutter, in the earth and the sky: the prayer of contemplation must of necessity lead to courageous and compassionate action. Prayer might begin with a sense of God beyond: "Our Father who art in heaven." But it dares to pray "thy Kingdom come" and moves to an awareness of the God nearby: "Thy will be done on earth as it is in heaven." This book aims to re-unite heaven and earth! Here we bring together ecology and spirituality: the climate of the planet illuminates the climate of the soul.

4 Invitation to Ecological Conversion

> *Most High, all-powerful, good Lord,*
> *Yours are the praises, the glory, and the honor, and all blessing . . .*
> *Praised be You, my Lord, with all Your creatures, especially Sir Brother Sun,*
> *Who is the day and through whom You give us light.*
> *And he is beautiful and radiant with great splendor;*
> *and bears a likeness of You, Most High One.*
> *Praised be You, my Lord, through Sister Moon and the stars,*
> *in heaven You formed them clear and precious and beautiful.*
> *Praised be You, my Lord, through Brother Wind,*
> *and through the air, cloudy and serene, and every kind of weather*
> *. . .*[23]

In these words, troubadour St Francis of Assisi invites us to recognize and celebrate the radical and essential interconnectedness of all things, displaying a remarkable kinship and sense of unity with creation in his *Canticle of Creation*, the first poem composed in the Italian vernacular. Hailing the sun as brother and the moon as sister, he greeted Sister Water and Brother Wind as dear friends. At the dawn of capitalism and a creeping consumerist approach to things—Francis was the son of a wealthy cloth-merchant and worked in his shop—he discovered a deep connectedness to all things which

---

22. Brown, *Spirituality and Liberation*.
23. Armstrong et al, *Early Documents*, 113–114.

was honoring and non-exploitative.[24] Pope Francis opens his 2015 encyclical *Care of our Common Home* with the words *Laudato Si*—"Praise be to you, my Lord"—quoting St Francis' *Canticle of Creation*. In his chapter "Ecological Education and Spirituality" Francis calls us to an ecological spirituality grounded in the convictions of our faith. Calling us towards a new lifestyle, he says that what we need is an "ecological conversion, whereby the effects of their encounter with Jesus Christ become evident in their relationship with the world around them. Living our vocation to be protectors of God's handiwork is essential to a life of virtue; it is not an optional or a secondary aspect of our Christian experience."[25]

Llewellyn Vaughan-Lee writes: "As the world grows more and more out of balance, we urgently need to regain a relationship with the planet based on the understanding of the world as a sacred living whole, and to reclaim a consciousness that is centered in that understanding."[26] May this book, as a barometer of the soul, enhance and develop your own consciousness, alertness and awareness to the world, outer and inner.

---

24. Franciscan prayer nurtures such an appreciative and respectful approach to the world of nature. See, for example, Stoutzenberger & Bohrer, *Praying with Francis*.

25. Pope Francis, *Laudato Si*, para. 217.

26. Vaughan-Lee, "The Call of the Earth" in Vaughan-Lee (ed.), *Spiritual Ecology*. For a scientific account of climate change see Klocek, *Climate*. For weather motifs in English literature, see Harris, *Weatherland*.

1

Mystic Metaphors
Reading the Soul

The heavens are telling the glory of God

(Ps 19:1)

COMMUNICATING THE MYSTERIES OF THE SOUL

How can we describe to others what is happening to us on our spiritual journey? How can we depict, for the benefit of ourselves and for others, the spiritual road that we are taking: experiences of prayer, transitions that we travel through, impediments that we face? Barry and Connelly put it: "most people are inarticulate when they try . . . to describe their deeper feelings and attitudes. They can be even less articulate when they try to describe their relationship with God. . . . For to begin to talk about this aspect of their lives requires the equivalent of a new language, the ability to articulate inner experience."[1]

---

1. Barry & Connolly, *Spiritual Direction*, 67.

Campbell tells us: "Here we sense the function of metaphor that allows us to make a journey we could not otherwise make."[2] As we seek to bring to expression aspects of our inner, spiritual life, we discover that we need metaphors, frameworks, reference points. Jurgen Moltmann affirms how vital it is to use images to describe spiritual experience: "In the mystical metaphors, the distance between a transcendent subject and its immanent work is ended . . . the divine and human are joined in an organic cohesion."[3]

Metaphors Enable us to be Spiritual Explorers

Paul Avis points out that metaphors drawn from the natural world are used by poets as a hermeneutical key to help map the landscapes of the mind: "Metaphor is generated in the drive to understand experience . . . Metaphor is not just naming one thing in terms of another, but seeing, experiencing and intellectualizing one thing in the light of the other."[4] Brian Wren puts it: "Metaphors can . . . extend language, generate new insights, and move us at a deep level by their appeal to the senses and imagination."[5] Metaphors have the power to shift us from left-brain analytical thinking to creative right-hemisphere imagining—and imaging.

In his study *The Edge of Words: God and the Habits of Language* Rowan Williams summons us to be unhesitating in our use of metaphors: "So as we take more risks and propose more innovations in our linguistic practice, we move from the more-or-less illustrative use of a vivid and unusual simile through to increasingly explosive usages that ultimately . . . invite us to re-think our metaphysical principles, our sense of how intelligible identities are constructed in and for our speaking. Extreme or apparently excessive speech is not an aberration in our speaking."[6]

Metaphors stoke and trigger the imagination. Janet Martin Soskice affirms their cognitive role, aiding and deepening our understanding of things: "what is said by the metaphor can be expressed adequately in no other way . . . the combination of parts in a metaphor can produce new and unique agents of meaning."[7] Metaphors evoke and stimulate rather than define or confine. Metaphors help to unify and integrate experience, because they link the spiritual to the physical, and the soul to the body, enabling the

---

2. Campbell, *Thou art That*, 9.
3. Moltmann, *Spirit of Life*, 285.
4. Avis, *God and the Creative Imagination*, 97.
5. Wren, *What Language Shall I Borrow?*, 92.
6. Williams, *Edge of Words*, 130.
7. Soskice, *Metaphor*, 31. See also Lakoff & Johnson, *Metaphors We Live By*.

metaphysical to become physical. There is a potential in metaphor properly described as *heuristic:* the word means "stimulating further investigation, encouraging discovery through experimenting, exploration something by first-hand experience." We need to rediscover the sacramentality of words: like bread and wine they can bear God's presence and reveal the Divine—so words should be approached with reverence and appreciation. Soskice affirms: "The sacred literature . . . both records the experiences of the past and provides the descriptive language by which any new experience may be interpreted."[8]

THE HEBREW SCRIPTURES DELIGHT IN CLIMATE METAPHORS

David's last words offer to us a divine description of true kingship drawn from the weather:

> The spirit of the Lord speaks through me,
>   his word is upon my tongue.
> The God of Israel has spoken,
>   the Rock of Israel has said to me:
> "One who rules over people justly,
>   ruling in the fear of God,
> is like the light of morning,
>   like the sun rising on a cloudless morning,
>   gleaming from the rain on the grassy land." (2 Sam 23: 2–4)

The prophet **Isaiah** also describes human activity in meteorological language:

> When the house of David heard that Aram had allied itself with
> Ephraim, the heart of Ahaz and the heart of his people shook as
> the trees of the forest shake before the wind. (7:2)
> Each will be like a hiding-place from the wind,
>   a covert from the tempest,
> like streams of water in a dry place,
>   like the shade of a great rock in a weary land. (32:2)

Chapter 25 describes both the divine and the human in meteorological terms:

> For you [O Lord] have been a refuge to the poor,
>   a refuge to the needy in their distress,
>   a shelter from the rainstorm and a shade from the heat.

---

8. Soskice, *Metaphor,* 160. See also Mayes, *Language of the Soul.*

> When the blast of the ruthless was like a winter rainstorm,
>   the noise of aliens like heat in a dry place,
> you subdued the heat with the shade of clouds;
>   the song of the ruthless was stilled. (25:4,5)

In chapter 35 the prophet alternates between poetic imagery drawn from the climate and more direct exhortations about human behavior:

> The wilderness and the dry land shall be glad,
>   the desert shall rejoice and blossom;
> like the crocus it shall blossom abundantly,
>   and rejoice with joy and singing . . .
> Strengthen the weak hands,
>   and make firm the feeble knees.
> Say to those who are of a fearful heart,
>   "Be strong, do not fear!
> Here is your God . . ."
> For waters shall break forth in the wilderness,
>   and streams in the desert;
> the burning sand shall become a pool,
>   and the thirsty ground springs of water. (Isa 35:1–4,6–7)

The prophet delights in natural imagery for the renewal of his people, but he also depicts Israel's enemies and those that threaten her in terms of the weather:

> The nations roar like the roaring of many waters,
>   but he will rebuke them, and they will flee far away,
> chased like chaff on the mountains before the wind
>   and whirling dust before the storm . . .
> This is the fate of those who despoil us,
>   and the lot of those who plunder us. (17:13,14)

Walter Brueggeman in his classic *The Prophetic Imagination* tells us that the role of the prophet is to envision an alternative consciousness, and to open up for people a different vision of things. The prophet enables an alternative perspective which may be subversive, questioning, compassionate, and which certainly reveals itself in counter-cultural lifestyle and political choices. So the prophets move from employing meteorological language of God, to using such imagery of human lives.

In Biblical literature the weather is often read metaphorically and evokes the movements of the soul. Changes in weather can help us describe and put into words what is happening in our inner life, while speaking to us of the Divine. **Hosea**, for example, gives us a dramatic juxtaposition of divine and human behaviors, described in imagery of climate:

> Let us know, let us press on to know the Lord;
>   his appearing is as sure as the dawn;
> he will come to us like the showers,
>   like the spring rains that water the earth.
> What shall I do with you, O Ephraim?
>   What shall I do with you, O Judah?
> Your love is like a morning cloud,
>   like the dew that goes away early. (Hos 6:3,4)

Hosea often describes human activity in such terms:

> Ephraim herds the wind,
>   and pursues the east wind all day long;
> they multiply falsehood and violence;
>   they make a treaty with Assyria,
>   and oil is carried to Egypt. (Hos 12:1)

> Sow for yourselves righteousness;
>   reap steadfast love;
>   break up your fallow ground;
> for it is time to seek the Lord,
>   that he may come and rain righteousness upon you. (Hos 10:12)

The psalmist greets "fire and hail, snow and frost, stormy wind fulfilling his command!" (Ps 148:8). He celebrates the message of the skies:

> The heavens are telling the glory of God;
>   and the firmament proclaims his handiwork. (Ps 19:1)

The psalms also describe the state of soul in weather terms. Psalm 72 offers this prayer for the king:

> May he live while the sun endures,
>   and as long as the moon, throughout all generations.
> May he be like rain that falls on the mown grass,
>   like showers that water the earth . . . .
> May his name endure for ever,
>   his fame continue as long as the sun.
> May all nations be blessed in him;
>   may they pronounce him happy. (Ps 72:5,6,17)

JESUS DELIGHTS IN CLIMATE METAPHORS

Jesus directs our attention heavenward in order to read the state of our soul. "He looked up to heaven" (John 17:1) in more than one sense. He says: "When it is evening, you say, 'It will be fair weather, for the sky is red.' And in the morning, 'It will be stormy today, for the sky is red and threatening.' You know how to interpret the appearance of the sky, but you cannot interpret the signs of the times" (Matt 16:2–3). He calls us to vigilance:

> But be alert; I have already told you everything . . .
> in those days, after that suffering,
> the sun will be darkened,
> and the moon will not give its light,
> and the stars will be falling from heaven,
> and the powers in the heavens will be shaken.
> Then they will see "the Son of Man coming in clouds" with great power and glory. Then he will send out the angels, and gather his elect from the four winds, from the ends of the earth to the ends of heaven. Beware, keep alert; for you do not know when the time will come. What I say to you I say to all: Keep awake. (Mark 13:23–27,33)

Directing our gaze towards the firmament, Jesus says: "Consider the birds of the sky" (Matt 6:26, CSB). "Consider": the Greek word means "turn your attention to this, notice what is happening, take a long, slow look, take note." Jesus summons us to a contemplative way of living, a deeply reflective way of seeing the world. Learn to see things differently. This sacramental approach to viewing reality becomes a dominant theme in the gospels, which combine to give us the clear impression that this was an outlook on the world that was truly characteristic of Jesus himself. The secrets of the Kingdom reveal themselves through parables of seed, mountain, field and sea (Matt 13, Mark 11:23). Jesus asks us to notice how the burning sun scorches the fragile shoots, how workers in the vineyard perspire and fatigue after a day under the unforgiving sun, how rain and wind batter down badly-founded homes. He asks us to "lift up your eyes" (John 4:35) and read what God is saying to us through the changing seasons.

In the gospels, then, we begin to notice the significance of climate metaphors. Indeed, they are found at the beginning and at the end of the ministry of Jesus. At his baptism, as Jesus emerges, dripping from the waters, the heavens are "torn apart" (Mark 1) evoking the reality that what has been closed and inaccessible is now open and unfolding; the opening of the heavens symbolizes the very nature and character of Jesus' ministry, which will witness the opening of a new way to God. At the transfiguration "his face shone like the sun" (Matt 17:2) and the dazzling light of the "bright cloud" (Matt 17:5) is symbolic of the illumination and enlightenment Jesus is giving to the disciples. Throughout the narrative the gospels offer us a symbolic universe and invite us to see reality more deeply. The physical storm on the lake is symbolic of the confusion and chaos in the disciples and mirrors their soul. (Later writers speak of us being "Buffeted by every wind of doctrine" as in Ephesians 4:14). At the crucifixion the accompanying earthquake and solar eclipse express the darkness of the soul and speak of the cosmic significance of the event. In a similar way, the sun rising at Easter daybreak becomes symbolic of a new beginning and a fresh dawn for humanity.

THEOLOGIANS AND MYSTICS CELEBRATE THE MACROCOSM OF THE SOUL

The Bible suggests that we can see the sky as the mirror of the soul. The King James Version gives us: "Hast thou with him spread out the sky, which is strong, and as a molten looking glass?" (Job 37:18). The NRSV offers us: "Can you, like him, spread out the skies, hard as a molten mirror?" For those with eyes to see it, the sky is the soul writ large. Theologians and mystics have recognized a potential microcosm/ macrocosm reciprocity—we can glimpse the movements of the soul in the heavens above us, and the skies help us read the movements of our own soul. There is an interplay between the atomic and the cosmic.

The Franciscan theologian **Bonaventure** (1221-74) invites us to develop such an outlook and way of seeing in his *Soul's Journey into God*. He detects clues and signs of the Divine throughout creation, which he calls the vestiges or divine fingerprints. For eyes that can see, elements in creation "are shadows, echoes and pictures of that first, most powerful, most wise and most perfect Principle . . . they are vestiges, representations, spectacles proposed to us and signs divinely given so that we can see God."[9] He invites us to activate our senses to celebrate and welcome the power, wisdom and goodness of God at every turn: "Therefore, open your eyes, alert the ears of your spirit, open your lips and apply your heart so that in all creation you may see, hear, praise, love and worship, glorify and honor your God."[10] Bonaventure invites us to notice both magnitude and minuscule details, hailing God in clouds overhead and in a dewdrop on the leaf. His contemporary **Thomas Aquinas** (1225-74) also conceived the human being as a

9. Cousins, *Bonaventure*, 26.
10. Cousins, *Bonaventure*, 30.

microcosm that encapsulates the entire cosmos by containing both spirituality and materiality, as the *imago Dei*, the image of God.

In the mystical tradition, outstanding writers and poets experience this reciprocity between heaven and earth. **Hildegard of Bingen** (1098–1179) writes:

> God has formed humanity according to the model of the firmament and strengthened human power with the might of the elements. God has firmly adapted the powers of the world to us so that we breathe, inhale, and exhale these powers like the sun, which illuminates the earth, sends forth its rays, and draws them back again to itself.[11]

She ponders how the human mystery mirrors the Divine, and how the soul is a microcosm of creation:

> In human beings there are body, soul and reason. The fact that I am aglow above the beauty of earthly realms has this meaning: the earth is the material out of which God forms human beings. The fact that I am illuminated in the water signifies the soul, which permeates the entire body just as water flows through the entire Earth. The fact that I am afire in the sun and the moon signifies reason: for the stars are countless words of reason.[12]

So we can read the soul by following clues in the heavens and earth. In one of Hildegard's visions God declares:

> I have created mirrors in which I consider all the wonders of my originality which will never cease. I have prepared for myself these mirror forms so that they may resonate in a song of praise ... This I have done, who am the Ancient of Days ... Humanity is the guise in which my Son, clothed in heavenly power, reveals himself as the God of all creation and as the Life of life.[13]

She offers us an invitation and a question:

> Glance at the sun.
> See the moon and the stars. Gaze at the beauty of earth's greenings.
> Now,
> think.
> What delight

---

11. Fox, *Illuminations*, 68.
12. Fox, *Divine Works*, 11.
13. Fox, *Divine Works*, 128.

> God gives
> to humankind
> with all these things.
> Who gives all these shining, wonderful gifts,
> if not God?
> Humankind should ponder God ...
> recognize God's wonders and signs ...
> The blowing wind,
> the mild, moist air,
> the exquisite greening of trees and grasses –
> In their beginning,
> in their ending,
> they give God their praise.[14]

Mechthild of Magdeburg (1210–82), in her work *The Flowing Light of the Godhead* gives us this exchange and reciprocity:

> *God says to the soul*
> O you beautiful sun in your radiance!
> O you full moon in the firmament!
> *The soul says to God*
> O you pouring God in your gift!
> O you flowing God in your love!
> O you burning God in your desire! ...
> You are a tempest in my heart.[15]

Looking Afresh at the World

As we begin our journey of exploration, we might learn from the astonishing sacramental world-view of **Ephrem** (306–373):

> In every place, if you look, His symbol is there. (Hymns on the Nativity 21:6)
> Lord, Your symbols are everywhere
> Blessed is the Hidden One shining out. (On Faith. 4:9)[16]

The Syriac tradition encourages us to read the two books of nature and Scripture. Ephrem affirms:

> The keys of doctrine

---

14. Uhlein, *Meditations with Hildegard*, 45–47.
15. Tobin, *Mechthild*, 48.
16. Brock, *Luminous Eye*, 55.

> which unlock all of Scripture's books,
> have opened up before my eyes
> the book of creation,
> the treasure house of the Ark,
> the crown of the Law.
> This is a book which, above its companions,
> has in its narrative
> made the Creator perceptible
> and transmitted His actions;
> it has envisioned all His craftsmanship,
> made manifest His works of art. (Hymns on Paradise 6:1,25)
> In his book Moses described the creation of the material world, so that both Nature and Scripture might bear witness to the creator:
> Nature, through man's use of it, Scripture, through his reading of it.
> These are the witnesses which reach everywhere,
> they are to be found at all times, present at every hour.
> (Hymns on Paradise 5:2)[17]

In this book we learn with Ephrem to see how the whole created order brims with the Divine and teaches us about God's ways. This is not a utilitarian approach to the natural world, looking around for helpful illustrations or analogies for the spiritual life. Rather, it is a question of training ourselves to recognize the revelatory character of creation and how God teaches us through it.

QUESTIONS FOR REFLECTION

1. As you begin the journey of this book, what are you seeking in your spiritual life?
2. What metaphors emerge when you begin to describe the course of your spiritual life over the last year?
3. What kind of metaphors do you tend to use most easily? Why is that, do you think?
4. What weather images would you use to express the state of your soul at this point? (Is your soul sunny, cloudy, chilly . . . ?)
5. What do you think can stimulate the art of wondering—reflection, musing—about God's presence in your life?

17. Brock, *Hymns on Paradise*, 108–109, 118, 102.

PRAYER EXERCISE

In contemporary times, with smart phones and ipads, we are often one step removed from the natural world, disconnected, literally losing touch with it. So explore with inquisitive eye your immediate environs. What do you notice? Celebrate the details of your surroundings, environment, setting, context.

The Franciscan Bonaventure (13C) encourages us to both appreciate the tiniest features on earth and also the magnitude and vastness of creation in sky and cloud:

> The beauty of things in the variety and light, shape and colour in simple mixed or organic bodies—such as heavenly bodies and minerals like stones and metals, and plants and animals—clearly proclaims the divine power, wisdom and goodness.

Recapture a sense of curiosity. Learn to be intrigued. Give thanks for the incarnate God, the God of the cosmos and the God of the detail. Ask yourself: where am I glimpsing God, or feeling his presence? What things speak to me of the Divine? Maybe read your environs symbolically and look for elements that somehow represent the sort of God you believe in. What metaphors of the soul emerge for you?

Perhaps you can sketch a picture or draw a symbol or write a poem to express your discoveries?

As you open up your awareness and consciousness of the Divine, conclude with thanksgiving.

2

Interpreting the Universe
Pilgrims, Prophets and Poets

*Blessed are you in the firmament of heaven,
and to be sung and glorified for ever . . .*
> *Bless the Lord, all rain and dew . . .*
> *Bless the Lord, all you winds . . .*
> *Bless the Lord, winter cold and summer heat . . .*
> *Bless the Lord, dews and falling snow . . .*
> *Bless the Lord, ice and cold . . .*
> *Bless the Lord, frosts and snows . . .*
> *Bless the Lord, lightnings and clouds*

sing praise to him and highly exalt him for ever.

(DAN 3:56–73)

THE DESTINY OF A PEOPLE—SHAPED BY THE ELEMENTS, FORMED BY THE CLIMATE

In this chapter we see how the climate and patterns of weather had a decisive effect on the story of the ancient people of the Israelites, effecting in the writings of the prophets the evolution of a symbolic universe, where the elements speak powerfully both of God and the soul. In the past century, biblical interpreters tended to oppose nature and history in their approach to the Hebrew scriptures, perceiving that the dominant theme was the story of redemption, with nature simply serving as a stage for the drama of salvation. Scholars such as Gerhard von Rad and G. Ernest Wright upheld this dichotomy, seeing biblical theology as emerging from the story not the context. Israelite religion was thus viewed as superior to and more sophisticated than the so-called "nature religions" of the time. But recently it has been appreciated that this divide is artificial and unhelpful: God's saving activity is discerned in the elements of the natural world. Creation—the heavens and the earth—are revelatory in their own right. Theophany occurs not only in the historical but also in the ecological. Hiebert, for example, affirms: "in antiquity, the forces of nature, which lay outside of human control but on which human survival absolutely depended, provided the most vivid images of the 'other', of the mystery at the heart of the universe . . . redemption is very much grounded in the world of creation."[1] This chapter whets the appetite for what is to come as we celebrate the diversity of weather references in all kinds of biblical material—historical/theological, poetry, song and wisdom literature.

Primordial Times

> In the beginning, God created everything: the heavens above and the earth below. Here's what happened: At first the earth lacked shape and was totally empty, and a dark fog draped over the deep while God's spirit-wind hovered over the surface of the empty waters. Then there was the voice of God "Let there be light." (Gen 1:1–2, *Voice*)

> On the day the heavens and earth were created, there were no plants or vegetation to cover the earth. The fields were barren and empty, because the Eternal God had not sent the rains to nourish the soil or anyone to tend it. In those days, a mist rose

1. Hiebert, *Yahwist's Landscape*, 151, 152.

up from the ground to blanket the earth, and its vapors irrigated the land. One day the Eternal God scooped dirt out of the ground, sculpted it into the shape we call human, breathed the breath that gives life into the nostrils of the human, and the human became a living soul. The Eternal God planted a garden in the east in Eden—a place of utter delight—and placed the man whom He had sculpted there. In this garden, He made the ground pregnant with life—bursting forth with nourishing food and luxuriant beauty. He created trees, and in the center of this garden of delights stood the tree of life and the tree of the knowledge of good and evil. A river flowed from Eden to irrigate the garden. (Gen 2:4–10, Voice)

The two accounts of creation celebrate in different ways the givenness and the gifting of the earth and its atmosphere. Indeed, the most typical form of Jewish prayer is the *berakhah*, sheer delight in the Divine, the blessing of God for his gifts: "Blessed are you, Lord God, King of the universe . . . " The watery elements are sheer gift, utter blessing, bespeaking a God who is unstoppable, unrelenting in his generosity and provision. But within a few paragraphs we read of the destructive powers of the element of water in the Great Flood:

> all of the subterranean waters erupted from the depths of the earth and burst skyward, covering the land. The casements of the heavens cracked open, dousing heavy rains over the watery earth for 40 days and 40 nights. (Gen 7:11,12, Voice)

The reader of the scriptures will quickly sense the paradoxical nature of the rains and waters: now life-giving, now destructive; now blessing, now judgment; now animating, now purging. The story of human origins and human downfall is told through the elements—there is no other way. Human destiny and the fate of the planet are inseparable and intertwined. In these opening verses of the Bible we already see how the elements can speak to the soul: of creation and redemption, of promise and judgment.

Further, in the narrative God declares the rainbow to be a symbol and sign of an everlasting covenant, an indicator and clue of God's faithfulness to all of humanity and to all creation:

> God said, "This is the sign of the covenant that I make between me and you and every living creature that is with you, for all future generations: I have set my bow in the clouds, and it shall be a sign of the covenant between me and the earth. When I bring clouds over the earth and the bow is seen in the clouds, I will remember my covenant that is between me and you and

every living creature of all flesh; and the waters shall never again become a flood to destroy all flesh. When the bow is in the clouds, I will see it and remember the everlasting covenant between God and every living creature of all flesh that is on the earth." (Gen 9:12–16)

We are forever to read significance and meaning into what appears in the sky.

Patriarchs

The story of the Hebrew people begins with **Abraham**, the ancestor and founding father of the Israelites. In promising him a new future, God directs Abram: "Look toward the heaven and number the stars, if you are able to number them.... So shall your offspring be" (Gen 15:5). He becomes the archetypal pilgrim. He is summoned to an adventure and journey of faith: "Now the Lord said to Abram, 'Go from your country and your kindred and your father's house to the land that I will show you.' They set forth to go to the land of Canaan" (Gen 12:1). The narrative goes on: "Abram journeyed on by stages towards the Negeb." But he does not settle there, not because of any hostile reception or threatening neighbors. He is impelled to move on precisely because of the weather, which will shape the very destiny of a people: "Now there was a famine in the land. So Abram went down to Egypt to reside there as an alien, for the famine was severe in the land" (Gen 12:9,10).

His great grandson **Joseph** will find himself retracing such a journey but under armed guard. His dream had alienated and infuriated his jealous brothers: "Look, I have had another dream: the sun, the moon, and eleven stars were bowing down to me" (Gen 37:9). They recognized themselves in these elements and conspired to have him kidnapped and taken into slavery. But in Egypt, Joseph interprets Pharaoh's dream: "The seven lean and ugly cows that came up after them [the good cows] are seven years, as are the seven empty ears blighted by the east wind. They are seven years of famine .... God has shown to Pharaoh what he is about to do" (Gen 41:27,28). Joseph's brothers duly arrive and plead with Pharaoh, "We have come to reside as aliens in the land; for there is no pasture for your servants' flocks because the famine is severe in the land of Canaan. Now, we ask you, let your servants settle in the land of Goshen" (Gen 47:4). Within a generation, the Hebrews found themselves enslaved as builders of pyramids, and so unfolds, under the leadership of **Moses**, the great formative epic of the Exodus.

Pilgrims

Within this drama, which becomes in Jewish and Christian thought the very paradigm of salvation and liberation, the elements of weather again play a central role. The "seven plagues of Egypt"—as they came to be known—were mainly weather-related, with specific details offered: "Moses stretched out his staff over the land of Egypt, and the Lord brought an east wind upon the land all that day and all that night; when morning came, the east wind had brought the locusts . . . The Lord changed the wind into a very strong west wind, which lifted the locusts and drove them into the Red Sea; not a single locust was left in all the country of Egypt" (Exod 10:13,19).

The winds are to play a decisive role in the great crossing of the Red Sea (or Sea of Reeds):

> The angel of God who was going before the Israelite army moved and went behind them; and the pillar of cloud moved from in front of them and took its place behind them. It came between the army of Egypt and the army of Israel. And so the cloud was there with the darkness, and it lit up the night; one did not come near the other all night. Then Moses stretched out his hand over the sea. The Lord drove the sea back by a strong east wind all night, and turned the sea into dry land; and the waters were divided. The Israelites went into the sea on dry ground, the waters forming a wall for them on their right and on their left. (Exod 14:19–22)

Moses' song celebrates the Divine in the elements of the Exodus crossing;

> At the blast of your nostrils the waters piled up,
>   the floods stood up in a heap;
>   the deeps congealed in the heart of the sea . . .
> You blew with your wind, the sea covered them;
>   they sank like lead in the mighty waters.
> Who is like you, O Lord, among the gods?
>   Who is like you, majestic in holiness,
>   awesome in splendor, doing wonders? (Exod 15: 8,10,11)

The very presence of God is, for the first time within this religious history, sensed in a dazzling cloud, amidst the howling winds. The wind of the Exodus is "the towering thunderhead typical of the Mediterranean storm front."[2] The divine cloud becomes a recurrent feature: in the desert, the Hebrews' craving for food is met by the gift of manna, "as Aaron spoke to the whole congregation of the Israelites, they looked towards the wilderness,

---

2. Hiebert, *Yahwist's Landscape*, 132.

and the glory of the Lord appeared in the cloud" (Exod 16:10). This is but a foretaste of the climatic and most significant theophany, on Mount Sinai:

> On the morning of the third day there was thunder and lightning, as well as a thick cloud on the mountain, and a blast of a trumpet so loud that all the people who were in the camp trembled. Moses brought the people out of the camp to meet God. (Exod 19:16,17)

As the journey through the wilderness progresses, Moses opens up for his people the glorious hope of a land flowing with milk and honey. There will be abundant rains from the heavens if the people live in undistracted worship to the Lord their God:

> If you will only heed his every commandment that I am commanding you today—loving the Lord your God, and serving him with all your heart and with all your soul— then he will give the rain for your land in its season, the early rain and the later rain, and you will gather in your grain, your wine, and your oil; and he will give grass in your fields for your livestock, and you will eat your fill. Take care, or you will be seduced into turning away, serving other gods and worshipping them, for then the anger of the Lord will be kindled against you and he will shut up the heavens, so that there will be no rain and the land will yield no fruit; then you will perish quickly from the good land that the Lord is giving you. (Deut 11:13–17)

Moses is emphatic:

> Behold, to the Lord your God belong heaven and the heaven of heavens, the earth with all that is in it; yet the Lord set his heart in love upon your fathers and chose their descendants after them, you above all peoples, as at this day. (Deut 10:14,15)

The epic journey of the Hebrews, and their arrival in the land of promise, are shaped by the visceral, physical experience of the elements, and the prophets of Israel will take this up as they strive to identify dramatic and powerful metaphors for God's continuing guidance and judgment. The elements enable the evolution of prophetic vocabulary, as the seers attempt to put into words humanity's greatest fears and hopes. The experience of diverse weather and changing climate will provide a lens—a hermeneutic key—enabling the prophets as they struggle to make sense of their people's ongoing failures and triumphs. It will be a key, indispensable ingredient in their reading of history and hope, a vital component as they evolve the language of the soul.

Prophets

The prophets lived in closest proximity to the earth and to the elements. They knew the land intimately, its contours and agriculture, the skies overhead. As they traversed valleys and canyons, and climbed hilltops, they discerned interconnections between land and sky, people and God.

Elijah reminds us how the prophets travelled the land and interacted with the unpredictable climate. Elijah ministered in the ninth century BC northern kingdom in time of Jezebel and Ahab. They were worshipping the Canaanite god Baal responsible for rain, thunder, lightning, and dew. Elijah warns Ahab that there will be years of catastrophic drought so severe that not even dew will form. On Mount Carmel he challenges the prophets of Baal and calls down fire from the heavens to consume his offering to the true God. After more than three years of drought and famine, God tells Elijah to return to Ahab and announce the end of the drought, presaged in clouds appearing over the Mediterranean coast. The withholding of rain is interpreted as a sign of divine disfavor; the return of rain is seen, not as a reward for repentance, but as a sign of the gracious faithfulness of the true God. Elijah retreats to Mount Horeb (1 Kgs 19), and later he is taken up into heaven in whirlwind (2 Kgs 2). His entire life had been lived in alertness to the elements. The unpredictable weather was not seen as bad luck or a misfortune, but as an indicator of divine displeasure. In the elements Elijah discovers clues and pointers to the Divine, which enable him to read the soul of his people and its leaders.

Amos the following century reveals a similar outlook. He is the first of the prophets to write down his message, which he clothes in dramatic imagery from the natural world. He says of himself: "I am no prophet, nor a prophet's son; but I am a herdsman, and a dresser of sycamore trees; the Lord took me from following the flock, and the Lord said to me, 'Go, prophesy to my people Israel'" (Amos 7:14,15). He lived close to the land in constant exposure to the elements. He discovers in them a language with which to communicate the divine will to the people. What is significant is that he passes naturally from the physical to the spiritual, from meteorology to spirituality. He can be talking about the literal elements:

> And I also withheld the rain from you
>   when there were still three months to the harvest;
> I would send rain on one city,
>   and send no rain on another city;
> one field would be rained upon,
>   and the field on which it did not rain withered;
> so two or three towns wandered to one town

> to drink water, and were not satisfied;
> yet you did not return to me,
> says the Lord. (Amos 4:7,8)
> Prepare to meet your God, O Israel!
> For lo, the one who forms the mountains, creates the wind,
>   reveals his thoughts to mortals,
> makes the morning darkness,
>   and treads on the heights of the earth—
> the Lord, the God of hosts, is his name! (Amos 4:12,13)

He turns his hearers to the heavens in wonderment and awe:

> The one who made the Pleiades and Orion,
>   and turns deep darkness into the morning,
>   and darkens the day into night,
> who calls for the waters of the sea,
>   and pours them out on the surface of the earth,
> the Lord is his name . . . (Amos 5:8)

But he goes on to pray that people will reflect God's faithfulness in their own human lives, for text continues:

> Seek good and not evil,
>   that you may live;
> and so the Lord, the God of hosts, will be with you. (Amos 5:14)

A little later he utilizes such imagery to powerfully convey God's longings:

> But let justice roll down like waters,
>   and righteousness like an ever-flowing stream. (Amos 5:24)
> The time is surely coming, says the Lord God,
>   when I will send a famine on the land;
> not a famine of bread, or a thirst for water,
>   but of hearing the words of the Lord. (Amos 8:11)

Streams from heaven speak to Amos of the need for justice, and he is anxious about his people's need for the sustenance of the divine words as well as for physical nourishment. One world provides clues for the other: Amos speaks of the deepest issues in the human heart by way of metaphors drawn from the natural world and from the very weather.

Jeremiah turns to the heavens to find vivid and dramatic contrasts with which to communicate his message. He contrasts the regularity of migratory birds in the sky with waywardness of his people:

> Even the stork in the heavens
>   knows its times;

> and the turtle-dove, swallow, and crane
> > observe the time of their coming;
>
> but my people do not know
> > the ordinance of the Lord (8:7)

He goes on to contrast useless idols with the creativity of the true God:

> Thus shall you say to them: The gods who did not make the heavens and the earth shall perish from the earth and from under the heavens.
> It is he who made the earth by his power,
> > who established the world by his wisdom,
> > and by his understanding stretched out the heavens.
>
> When he utters his voice, there is a tumult of waters in the heavens,
> > and he makes the mist rise from the ends of the earth.
>
> He makes lightnings for the rain,
> > and he brings out the wind from his storehouses.
>
> (Jer 10:11–13)

Poets

In his major study *Weathering the Psalms,* Steve A. Wiggins affirms: "the conviction that weather touches the very heart of ancient perceptions of the Divine . . . To understand the weather is somehow to glimpse the Divine."[3] As the Psalmist puts it: "you make the winds your messengers." Psalm 65 delights in the providence of God:

> You visit the earth and water it,
> > you greatly enrich it;
>
> the river of God is full of water;
> > you provide the people with grain,
> > for so you have prepared it.
>
> You water its furrows abundantly,
> > settling its ridges,
>
> softening it with showers,
> > and blessing its growth.
>
> You crown the year with your bounty;
> > your wagon tracks overflow with richness.
>
> The pastures of the wilderness overflow,
> > the hills gird themselves with joy,
> > the meadows clothe themselves with flocks,

---

3. Wiggins, *Weathering the Psalms,* 3.

> the valleys deck themselves with grain,
> they shout and sing together for joy. (Ps 65:9–13)

Psalm 8 marvels at the dignity of human beings, who might seem so insignificant in the face of wide skies and unfrontiered space:

> When I look at your heavens, the work of your fingers,
> the moon and the stars that you have established;
> what are human beings that you are mindful of them,
> mortals that you care for them? (Ps 8:3,4)

Sages

Such questions lead us to the wisdom literature in the Hebrew scriptures. What is striking in the wisdom literature is that weather references are almost entirely describing human behaviors, not divine. See these incisive references to rain in the **Book of Proverbs**:

> In the light of a king's face there is life, and his favor is like the clouds that bring the spring rain. (16:15)
> A stupid child is ruin to a father, and a wife's quarrelling is a continual dripping of rain. (19:13)
> Like snow in summer or rain in harvest, so honor is not fitting for a fool. (26:1)

In similar vein, the wind points not to the divine Spirit, but to human conduct:

> Those who trouble their households will inherit wind, and the fool will be servant to the wise. (11:29).
> Like clouds and wind without rain is one who boasts of a gift never given. (25:14)
> The north wind produces rain, and a backbiting tongue, angry looks. (25:23)
> . . . to restrain her is to restrain the wind or to grasp oil in the right hand. (27:16)

The Book of Ecclesiastes opens with this observation of regularity, even predictability:

> The sun rises and the sun goes down, and hurries to the place where it rises.
> The wind blows to the south, and goes round to the north;

round and round goes the wind, and on its circuits the wind returns.
All streams run to the sea, but the sea is not full;
to the place where the streams flow, there they continue to flow. (1:5–7)

Ecclesiastes gives us thirty two references to sun, with "under the sun" the oft-repeated refrain. Fourteen times the author refers to wind, mainly "chasing after wind," an evocative phrase denoting emptiness or unfulfillment: "I saw all the deeds that are done under the sun; and see, all is vanity and a chasing after wind." His closing advice is: "Remember your creator in the days of your youth, before the days of trouble come, and the years draw near when you will say, 'I have no pleasure in them'; before the sun and the light and the moon and the stars are darkened and the clouds return with the rain" (12:1,2).

This brief survey prepares us for a more thorough exploration of the elements. We are beginning to see that weather references play a key role in the unfolding story of God's people. We recognize too that biblical writers turn to the heavens both for pictures of the Divine and for metaphors of the human.

QUESTIONS FOR REFLECTION

The different types of Scripture pose their own questions:

1. *The Historical books* challenge us to discern the action of God in history and in the present. As interpretations of events within a narrative-theology that seeks to make sense of the experience and vicissitudes of God's people, they invite us to identify and name the work of God in today's world. What is going on, and what is God up to today? What meteorological metaphors come to mind, as you ponder this?

2. *The Prophetic books* present a special challenge to ministry today. They invite us to uncover hypocrisy and idolatry, and to recall God's people to courageously struggle for peace and justice and the true knowledge of God. What metaphors from the prophets inspire you most?

3. *The Psalms* and poetic works have a cherished place in ministry today, as we identify with the hopes and hurts of the psalmist. They teach us not to hesitate in bringing to God in prayer our questions and our doubts. What do they suggest to you, in terms of the climate of the soul?

4. *The Wisdom literature* invites us to make sense of today's confusions by pondering anew the meaning of a life well-lived, with the providence of God. Re-reading the references from Proverbs, how would you express key convictions about life—through weather metaphors?

PRAYER EXERCISE

Take a psalm celebrating the climate (for example, Pss 65, 148) and pray with it. Write your own psalm or poem, expressing the themes of gratitude and hopefulness.

3

Facing the Elements
Jesus' Experience and Message

He who descended is the same one who ascended far above all the heavens, so that he might fill all things

(EPH 4:10)

OUT IN THE OPEN

"The Son of Man has nowhere to lay his head" (Luke 9:58). Jesus himself was exposed to staggering contrasts in environmental conditions, and stunning extremes of temperature, from fierce heat to bitter, biting cold. He traversed the land, moving from the sizzling, scorching desert to the balmy sub-tropical area of the Sea of Galilee (Mark 1:12–14), travelling from the breezy and windswept coast and beaches of Tyre and Sidon to the mountains of the Golan Heights (Mark 7:31), before approaching snow-capped Mount Hermon by Caesarea Philippi (Mark 9:2). In a Jerusalem set high in the hills he takes shelter in the Temple precincts in winter (John 10:22,23) and feels the chill in his vigil in Gethsemane (John 18:18). Jesus was often out in the open and subject to the vicissitudes of the weather. Ever on the road as pilgrim, wayfarer and itinerant preacher, his sweating forehead did not escape the blazing sun; he felt the variable winds sweeping through his hair and caressing his burnished skin, the rain drenching his clothes. The gospel narratives sometimes refer to the conditions of sky: "That evening, at sunset, they brought to him all who were sick or possessed with demons . . . In the morning, while it was still very dark, he got up and went out to a deserted place, and there he prayed" (Mark 1:32,35 cf. Luke 4:40). Celebrating the incarnation, Christians do not believe in a distant God, locked up in a faraway heaven, an other-worldly deity. They believe in a sweaty, dusty God revealed in the humanity of Jesus, vulnerable to the elements, and sharing our human condition utterly and completely. But paradoxically, the skies above us speak of the otherness of God, God's utter transcendence, the "Beyond." We glimpse the Divine both in the wind-swept storm-tossed humanity of Jesus and in the wild elements that sweep across the heavens.

The Kingdom of the Heavens

As Matthew's gospel gives it to us, Jesus' message was "The kingdom of the heavens" (Matt 3:2, the Greek is plural). In a sense, Jesus wants to lead us from earth to heaven, and directs our gaze skywards. Here we will note some key sayings by Jesus about the climatic conditions—we will take a closer look at some of them later on.

> When you see a cloud rising in the west, you immediately say, "It is going to rain"; and so it happens. And when you see the south wind blowing, you say, "There will be scorching heat"; and it happens. You hypocrites! You know how to interpret the appearance of earth and sky, but why do you not know how to interpret the present time? (Luke 12:54–56)
> The Pharisees and Sadducees came, and to test Jesus they asked him to show them a sign from heaven. He answered them, "When it is evening, you say, 'It will be fair weather, for the sky is red.' And in the morning, 'It will be stormy today, for the sky is red and threatening.' You know how to interpret the appearance of the sky, but you cannot interpret the signs of the times." (Matt 16:1–3)

Jesus does not despise a looking to the sky, but he wants this to lead to a looking to the soul, an attentiveness to what is going on in society. Jesus refers to the mystery and majesty of the wind:

> The wind blows where it chooses, and you hear the sound of it, but you do not know where it comes from or where it goes. So it is with everyone who is born of the Spirit. (John 3:8)

> As they went away, Jesus began to speak to the crowds about John: "What did you go out into the wilderness to look at? A reed shaken by the wind?" (Matt 11:7).

At the beginning of his ministry, in Nazareth, Jesus recalls the drought: "But the truth is, there were many widows in Israel in the time of Elijah, when the heaven was shut up for three years and six months, and there was a severe famine over all the land" (Luke 4:25). In the Sermon on the Mount the sun and the rain have their message, too, pointing back to our relationships here on earth:

> I say to you, Love your enemies and pray for those who persecute you, so that you may be children of your Father in heaven; for he makes his sun rise on the evil and on the good, and sends rain on the righteous and on the unrighteous. (Matt 5:44,45)

The Parables

There are significant references to the elements in the parables. Jesus experienced for himself the singeing power of the sun's rays when in the desert, and twice he refers to the blistering burning sun. In the parable of workers in the vineyard the laborers complain: "These last worked only one hour, and you have made them equal to us who have borne the burden of the day and the scorching heat" (Matt 20:12). In the parable of the sower the sun withers vulnerable new shoots (Matt 13:3–9). This is given a symbolic meaning in the interpretation of the evangelist, the scorching sun coming to represent difficulty or persecution: "As for what was sown on rocky ground, this is the one who hears the word and immediately receives it with joy; yet such a person has no root, but endures only for a while, and when trouble or persecution arises on account of the word, that person immediately falls away" (13:21). This is echoed in the parable of the builder, where the onslaught of the elements represents things that threaten obedience to Jesus' word: "Everyone then who hears these words of mine and acts on them will be like a wise man who built his house on rock. The rain fell, the floods came, and the winds blew and beat on that house, but it did not fall, because it had been founded on rock" (Matt 7:24–25).

The Close of the Age

In Matthew 24:27–35 he makes six references in succession to the elements:

> For as the lightning comes from the east and flashes as far as the west, so will be the coming of the Son of Man.
> The sun will be darkened,
>   and the moon will not give its light;
> the stars will fall from heaven,
>   and the powers of heaven will be shaken.
> Then the sign of the Son of Man will appear in heaven, and then all the tribes of the earth will mourn, and they will see "the Son of Man coming on the clouds of heaven" with power and great glory. And he will send out his angels with a loud trumpet call, and they will gather his elect from the four winds, from one end of heaven to the other.
> From the fig tree learn its lesson: as soon as its branch becomes tender and puts forth its leaves, you know that summer is near. So also, when you see all these things, you know that he is near, at the very gates. Truly I tell you, this generation will not pass away until all these things have taken place. Heaven and earth will pass away, but my words will not pass away.

KEY MOMENTS DENOTED BY THE WEATHER

The gospel narratives are punctuated by references to the sky. We should note that these come at the most significant and poignant moments. At the very outset, at the baptism of Jesus, closed heavens are torn wide open (Mark 1:10). At the transfiguration dazzling sunlight passes to impenetrable cloud. On Palm Sunday there is thunder (John 12:29). At the crucifixion, there is an eclipse of the sun. As we will consider the chapter six, the weather-related events of the crucifixion serve to link the Cross with the cosmos, including its pain and chaos. Matthew in particular, who dramatizes his account with earthquakes at the crucifixion and resurrection, as well as including the solar eclipse, wishes the reader to link these events with the primordial theophany of Mount Sinai, where God appears amidst cloud, fire and thunder. Such imagery echoes Jesus' reference to "but the beginning of the birth pangs" (Matt 24:8) and anticipates Paul's cosmic vision of the new creation emerging from the groaning in travail amidst the whole creation (Rom 8:18–24). In the thought of the Letter to the Hebrews, Sinai of old points to Mount Zion to come: "You have not come to something that can be touched, a blazing fire, and darkness, and gloom, and a tempest . . . But you have come to Mount Zion and to the city of the living God, the heavenly Jerusalem . . . to Jesus, the mediator of a new covenant" (Heb 12:18–24).

Indeed, in the ascension, Jesus is received into the cloud (Acts 1:9). As we shall see in the next chapter, John's gospel reflects the rhythm of the passing seasons and highlights the annual Passover which shapes the very narrative.

The Storm on the Lake

In all four gospels we encounter the storm on the lake:

> On that day, when evening had come, he said to them, "Let us go across to the other side." And leaving the crowd behind, they took him with them in the boat, just as he was. Other boats were with him. A great gale arose, and the waves beat into the boat, so that the boat was already being swamped. But he was in the stern, asleep on the cushion; and they woke him up and said to him, "Teacher, do you not care that we are perishing?" He woke up and rebuked the wind, and said to the sea, "Peace! Be still!" Then the wind ceased, and there was a dead calm. He said to them, "Why are you afraid? Have you still no faith?" And they were filled with great awe and said to one another, "Who then is this, that even the wind and the sea obey him?" (Mark 4:35–41)

Jesus issues a double summons, which echoes across the centuries: "Put out into the deep!" (Luke 5:4); "Let us go across to the other side" (Mark 4:35). It seems significant that Jesus first calls his disciples by the shoreline. Standing on the strand, the choppy waters of Galilee lapping at his feet and stretching out before him, Jesus says to the fishermen: "Follow me!" (Mark 1:17). He wants to lead them away from the safety of the bank, the homes of Capernaum which represent their place of comfort and retreat[1] to uncharted waters exposed to the elements. He wants to lead them from enclosure to exposure, from security to vulnerability. At one point we read: "He told his disciples to have a boat ready for him" (Mark 3:9). Jesus has challenging journeys in store for his disciples . . .

The awesome, untamable waters of the Lake summon the disciples to risk and adventure. When we read the account of the stilling of the storm in the gospels we often interpret the event in terms of a miracle: it denotes the violation or suspending of the laws of nature. Such an approach derives from the post-enlightenment worldview. In his philosophy David Hume defined "miracle" as a violation of nature because he worked with a cosmology—which he shared with Newton—which saw the universe as a closed,

---

[1]. The house is a particular feature in Matthew's gospel. See Crosby, *House of Disciples*. See also *Pennington, Heaven and Earth*.

predictable, mechanistic system, and when the universe is seen as a machine, new things just don't happen. Our post-modern cosmology is much more open, fluid and dynamic. In the gospels, events like the calming of the storm are not called miracles—the word used is either *dumanis*, a work of mighty power, or *semeia*, as in John's gospel, a sign, that points to God's revelation. What worldview, then shaped the authors and first readers of the gospels?

In the Ancient Near East, the sea was feared as the abode of chaos. The waters are brimming with demons and monsters. We even know their names! Here lurks Leviathan and Rahab (Ps 74:13–14). For the first readers of the Gospels, the very mention of the sea evokes the primordial chaos of Genesis 1: "darkness covered the face of the deep, while a wind from God swept over the waters." The Lake also looks forward to the End: the defeat of the beast in the sea, as in Daniel and Revelation. (Indeed, at the End, Rev 21:1, "there is no more sea"). This is all very much part of the first century thought world. Such a mindset can be traced across the Ancient Middle East: the Babylonian Epic of Gilgamesh conceives the waters as a place of danger and terror, while the Ugarit texts narrate how Baal battles with the sea-god Yam. Psalm 107 observes the dangers sailors faced:

> They mounted up to heaven, they went down to the depths;
>   their courage melted away in their calamity;
> they reeled and staggered like drunkards,
>   and were at their wits' end.
> Then they cried to the LORD in their trouble,
>   and he brought them out from their distress;
> he made the storm be still,
>   and the waves of the sea were hushed.
> Then they were glad because they had quiet,
>   and he brought them to their desired haven. (Ps 107:23–30)

In Mark 4, in the raging midst of the storm, Jesus is asleep on a cushion (4:38): an image of perfect trust, constancy, stability and resting in God's providence and in God's purposes. The command that Jesus issues to the elements of the storm, which we normally translate as "Peace, be still!" is literally, in the Greek *(pephimoso)*, "be muzzled!" This is the same command Jesus issues to the demon in the synagogue at Capernaum (Mark 1:25). Jesus is addressing the demons of the deep.

The key question is; "Who is this, that even the sea and wind obey him?" Mark's gospel gives us the answer that this is no less than the Lord of Creation—the one who brings order into chaos, the one who defeats evil, restores creation and inaugurates the very renewal of the earth. It is to the great theme of the Cosmic Christ that we will turn later.

For the moment, notice this: in all Jesus' own references to weather and climate he is turning his disciples outwards in their focus to observe what is going on in the wider world. The episode of the storm on the lake, from the earliest days of Christian reflection, has been interpreted as alluding both to the personal storms we face and the storms encountered in the church and in society. In the midst of the storm Jesus speaks out directly to the negative destructive forces: the addresses the demons and dragons of the deep. He also attends to the anxiety and inner storms of the disciples, asking "why are you afraid?" We conclude this chapter looking at these storms inner and outer.

Storms of the Spirit

The same lake that experienced the storm also witnessed Jesus seeking retreat for his disciples:

> The apostles gathered around Jesus, and told him all that they had done and taught. He said to them, "Come away to a deserted place all by yourselves and rest a while." For many were coming and going, and they had no leisure even to eat. And they went away in the boat to a deserted place by themselves. (Mark 6:30,31)

The disciples had been in a whirlwind of activity, and Jesus himself had experienced a storm of protest and rejection in Nazareth (Mark 6:1–6). Now it was time to find a healing calm in the *eremos*—the solitary place of retreat. Now the boat is redirected.

Paul lists as one of the gifts of the Spirit *kubernetes*: lamely translated "administration," it means "navigation" or "helmsmanship" (1 Cor 12:28). For Paul, the art of discerning the Spirit's movement, the art of recognizing the need of the moment, is akin to the skill of the ship's pilot and steersman who, working collaboratively alongside the captain, coxswain, and the entire crew, will guide the ship in its adventures. In the seventh century John Climacus of Sinai speaks of the need for courageous spiritual directors using this image: "A ship with a good navigator comes safely to port, God willing."[2]

Sometimes the spiritual life feels as if we find ourselves in the midst of a tempest or whirlwind. We are buffeted to and fro. We feel insecure and

---

2. Luibheid & Russell, *Climacus*, 259.

unsure of ourselves. We might join the evangelical Priscilla J Owens in the words of her 1882 hymn:

> *We have an anchor that keeps the soul*
> *Steadfast and sure while the billows roll,*
> *Fastened to the Rock which cannot move,*
> *Grounded firm and deep in the Savior's love.*
> *It is safely moored, 'twill the storm withstand,*
> *For 'tis well secured by the Savior's hand;*
> *And the cables, passed from His heart to mine,*
> *Can defy that blast, thro' strength divine.*

What is your own experience of facing the storm of prayer? Who or what guided you through this experience, and how might it help you in supporting others?

Storming the Institution: Prophetic Spirituality

In his ministry Jesus encounters many conflicts and struggles, not least the storm of controversy: "He looked around at them with anger; he was grieved at their hardness of heart" (Mark 3:5). He enters into the fray with the Pharisees and Sadducees, for example, in Matthew 23. Jesus storms into the Temple in the event sometimes called the "cleansing of the Temple." Quoting Jeremiah and Isaiah he locates himself in the tradition of Jerusalem prophets (Isa 56:7, Jer 7:11). This event is best interpreted as a prophetic action declaring in word and deed the ending of the Temple worship.

Jesus had often been recognized as a prophet. At Nain, after Jesus has spoken words of hope to a grieving widow and lifted up her dead son to new life, the people "glorified God, saying, 'A great prophet has risen among us!'" (Luke 7:16). Jacob's Well witnesses the woman's acclamation: "Sir, I see that you are a prophet" (John 4:19). In Jerusalem, he is hailed in prophetic terms: "So they said again to the blind man, 'What do you say about him? It was your eyes he opened.' He said, 'He is a prophet'" (John 9:17). At the Palm Sunday entry into Jerusalem the crowds hailed him: "This is the prophet Jesus from Nazareth in Galilee" (Matt 21:11). Soon afterwards, we read, "They wanted to arrest him, but they feared the crowds, because they regarded him as a prophet" (Matt 21:46). In the fourth gospel, Jesus is twice acclaimed a prophet in a context about political choices. "When the people saw the sign that he had done, they began to say, 'This is indeed the prophet who is to come into the world'" (John 6:14). There was a division among the people:

"When they heard these words, some in the crowd said, 'This is really the prophet'" (John 7:40).

What is a prophet? The prophets of old both spoke the word of God and also embodied or symbolized the word in a dramatic action. The burden of the Hebrew prophets was not prediction of the future, but rather declaring God's word into the present situation, naming the idols and illusions of contemporary society.[3] When Jesus overturns the money-changers' tables in the Temple precincts he is indeed storming—disrupting the status quo and confronting head-on those who represent the powers-that-be. No gentle Jesus meek and mild here! In art, hands that had healed the sick become clenched fists raised in righteous anger.[4] Kicking over the stalls and knocking aside the seats, he goes further: "Making a whip of cords, he drove all of them out of the Temple, both the sheep and the cattle. He also poured out the coins of the money-changers" (John 2:15). He speaks harshly and penetratingly to the traders. It is stormy weather in Jerusalem!

Most of us want to avoid conflict at all costs. But the world confronts us with many instances of injustice, exploitation and hypocrisy, raising uncomfortable questions about our own spirituality and prayer. Are there danger signs that my spirituality is becoming narcissistic, self-centered, closed in on itself? Is my spirituality about self-fulfillment or about empowering sacrificial living? If the measure of spiritual maturity is increasing solidarity with the hurting, an enlarging capacity for compassion, what are the signs that I am maturing? Is my heart getting bigger? How far can I allow the pain of the world to enter my prayer? Does my prayer have room for the oppressions and injustices of the world? How can I stand in solidarity with those facing stressful storms across the globe? What place is there for a costly intercession which is inseparable from self-offering (and does not let me "off the hook")? Indeed, what is my understanding of intercession? Is it advising the Almighty or "coming before God with the people on your heart" (Michael Ramsey)? What place is there in my prayer for the Cross—not only in terms of seeking personal forgiveness but in realizing that God suffers among us? What does Matthew 25 look like in my experience? What is the evidence? Am I drawn to the margins in any way? As Jim Wallis puts it: "Personal piety has become an end in itself instead of the energy for social justice . . . Prophetic spirituality will always fundamentally challenge the system at its roots and offer genuine alternatives based on values from our truest religious, cultural and political traditions."[5]

---

3. See, for example, Amos 5:24; Isa 58:6,7.
4. For example, as in Giotto's fresco in the Scrovegni Chapel, Padua.
5. Wallis, *Soul of Politics*, 38, 47.

Storms in the Soul: Jesus faces Inner Disturbance

With something like the sound of thunder rolling overhead in darkening Jerusalem skies, Jesus cries out: "Now my soul is troubled. And what should I say—'Father, save me from this hour'?" Storms overhead mirror the storm gathering in Jesus' soul (John 12:27,29). Just previously, approaching the tomb of Lazarus, the evangelist tells us "he was greatly disturbed in spirit and deeply moved . . . Jesus looked upwards" (John 11:33, 41). Recounting the events of Gethsemane, Luke describes Jesus as being in an *agonia* (14:44). This is the only time the word is used in the New Testament, and carries the meaning of an intense struggle for victory, wrestling, contest: Gethsemane becomes a battlefield with the powers of darkness. The word denotes both physical and spiritual suffering. "He began to be distressed and agitated. And he said to them, 'I am deeply grieved, even to death'" (Mark 14:33,34). Mark uses here three powerful verbs which take us into the turbulence of Jesus' inner storms. *Ekthambeisthai* means he became utterly dismayed, tormented. It can also have the sense of being astonished, astounded. *Adeemonein* conveys the sense that he was seized by horror, gripped by a very intense anguish. *Perilupos* means he is grieved, exceedingly sorrowful, heartbroken, overcome with sadness. The heavy darkness presses Jesus into the dust of the ground. The soil is wet, not with the morning dew, but with the tears of Christ: "Jesus offered up prayers and supplications, with loud cries and tears, to the one who was able to save him from death" (Heb 5:7). What is going on here? And how does it resonate with our prayer?

Jesus is standing in solidarity with all those who face their darkest hour. In this most human of scenes, Jesus is identifying with all those who through time will be crushed by their destiny and find themselves prostrate on the ground. But one fact is significant above all else. He brings his questions, doubts and fears to one he calls *Abba*. He exposes his heart's anxieties to *Abba*. In the darkness of the garden, he enters into an intimacy with the Father, and, as a result of this, Gethsemane becomes a threshold, a place of crossing. There seems to be a triple movement involved.

First, there is a passage from attachment to bereavement. Secondly, Jesus transits from terror to trust, from agitation to composure. Thirdly, Jesus makes a passage from resistance to surrender. Slowly yet deliberately and decisively, there is an abandonment to God, a yielding, a submission, a giving in, a movement from hesitant holding back to courageous self-emptying.

Within a few days, a radiant Jesus will be addressing his disciples with the words: "Peace be with you!" (John 20:19). But first he must face, in the passion, the most turbulent hours of his life. The storms he embraces, of course, turn out to be the most transformative events for humanity!

QUESTIONS FOR REFLECTION

1. Looking again at Jesus' own references to weather and climate, how can you sharpen and hone your skills of looking at the world sacramentally, noticing symbolism and messages in the environment?
2. What would your life look like if you modeled it on Jesus the prophet? When was the last time you were outraged or indignant about an injustice, near or far, and actually did something about it?
3. The ministry of the prophet is characterized by courage and audacity, by plain speaking and bold actions. How do these reveal themselves in your life?
4. Dare you be disruptive in relation to society's status quo? Do you play it safe and stay out of trouble? Is your life predictable or tame? Do you need to recover nerve and verve?
5. What would a prophetic spirituality look like in your context? What storms dare you enter?

PRAYER EXERCISE

Either

Revisit the story of the disciples on the lake at night, Matthew 14:22–33. There are 8 stages of this prayer exercise as we follow through the gospel account.

1 Embarking

You and the disciples set sail. Jesus tells them to. *He made the disciples get into the boat and go on head to the other side . . . (v 22)* Climb into the boat. What do you notice as you look around? It is evening: is it dark already? Are the waters inviting or threatening? Visualize the scene, as vividly as you can. What are the other disciples thinking? Listen to their chatter. Are they excited or nervous? What are you feeling as the journey begins? Use your five senses. What can you see? Describe the landscape. What can you smell? Can you taste the breeze? What sounds can you hear? What are you touching?

2 Storm

when evening came ... the boat, battered by the waves, was far from the land, for the wind was against them (v 24)
Can you feel the wind blowing through your hair and clothes? Is the sea inky black yet, or still darkening from sunset pink to marine blue? Do you feel safe as the waters turn choppy, thrashing the side of the boat, and as the wind gets up? How are you feeling right now?

3 Fear

You see a vague outline on the surface of the deep: it looks like a man, but it can't be. *But when the disciples saw him walking on the sea, they were terrified, saying, "It is a ghost!" (v 26)*. Do you feel a spine-chillingly sensation, the hairs of your head as it were standing on end? Can you hear your heart thumping? What does it feel like to be afraid? Is there sweat in the palms of your hand? What fears are crippling you right now. Name your fears. Then, when you are ready, hear his voice: *"Take heart, it is I, do not be afraid!"* How does that make you feel?

4 Desire

Peter answered him, "Lord, if it is you, command me to come to you on the water" (v.28)
What do you want most, right now? What is your heart's desire, your deepest longing, your greatest need, at this point in your life? As Jesus said to the blind man, so he says to you: "what do you want me to do for you?" What is your honest response to this question, from the depths of your heart? No one else will hear you! Say it!

5 Invitation

You feel strangely and inexplicably impelled to go to Jesus. But you know this is suicidal, to climb out of a boat in a dark storm. Jesus says to you "Come." You love the boat. It is yours, you own it. It is where you are in charge, where you like to be, normally. It is where you give orders to the other fishermen. This is your place of safety. But you find yourself moving to the edge of the boat. You are putting one leg over the side!

6 Solitude

Peter got out of the boat (v. 29). As Peter does this action entirely on his own bat, so you feel now very alone. It is just you and Jesus. The others fade from view, and their views hardly count now. You see the blackness. You feel the howling wind. Your foot feels the uncertain, heaving surface. As the waters rise and fall, rise and fall, what is the paradox you are wrestling with? What dragons lurk under the waters? What is the sign of contradiction in your life? What is coming up to the surface in your feelings? Name your questions. Name your emotions right now. As the paradox moves you backwards and forwards, to and fro, expose these to the wind, the gale of the Spirit upon the face of the waters. Feel the Spirit- breath whistling in your ears and the divine waves pounding upon your soul. Feel the energy. Let the place of paradox be a vibrant, creative place. Realize you don't have to solve a paradox. You live it. It may be feel a risky place. But you are learning to walk on water! The God of the impossible (Luke 1:37) is upholding you. For as long as you can, stay with Jesus in the darkness. He is not faraway. He is within reach. But you are standing alone before him and the raging elements.

7 Handclasp

As you feel you are sinking, Jesus reaches out his hand to you. He holds you. You will be OK. As he had a question for Peter *("why did you doubt?," v 31)*, what is he asking of you? What question do you hear from Jesus?

8 Response

Those in the boat worshipped him (v. 33). Getting back in the boat, you feel relief that you are, in a sense at least, "back home," out of danger. What do you want to say to Jesus? Conclude by giving thanks. Then let the boat take you safely back to shore. You will never be the same again! *When they had crossed over, they came to land at Gennesaret (v. 34)*
What do you want to take away from this, and maybe discuss with a spiritual director or soul friend?

Or

Use your hands expressively in this prayer-time in four actions.

Begin by clenching your fists tight and holding them before you. Feel the tension and let these fists represent an anger or frustration that bothers you today, a situation in the world that you feel strongly about. Hold them before God in the solidarity of prayer and intercession.

Secondly, slowly open your down-turned palms and let go of the tension. Let it fall away from you to God. In this gesture give to God any negative feelings or stresses, feel them drip out of your fingertips, as it were. Surrender the situation to God's providence and sovereignty.

Thirdly, turn your hands upwards in a gesture of surrender to God and of receiving from God. Breathe in what God wants to give you right now—perhaps a reassurance that all will be well. Breathe in his empowering Spirit who will give you the courage for action.

Finally, take a look at your hands. Is there an action that God is calling you to make in relation to your initial concern? What should you do as a result of this—something bold, something risky or rebellious?

Recall 2 Timothy 1:7 "God did not give us a spirit of cowardice, but rather a spirit of power and of love and of self-discipline." End with the Serenity Prayer: "God grant me the serenity to accept the things I cannot change; courage to change the things I can; and wisdom to know the difference."

4

Welcoming the Unpredictable
Seasons of the Spirit

As long as the earth endures,
  seedtime and harvest, cold and heat,
summer and winter, day and night,
  shall not cease.

(GEN 8:22)

Weather shapes our lives and has an unrelenting influence, even on our worship and prayer. The very worship in Jerusalem's Temple was shaped around the pilgrim festivals that originated as agricultural celebrations marking the annual shifts in climate and weather. In this way the rhythms of the climate were ritualized and celebrated in song and dance and sacrifice. Autumn saw the fruit harvest and laborers built shelters in the orchards during the picking season—hence, the feast of tabernacles or *Sukkoth* arose.[1] The darkness of winter is dispelled by the festival of lights—*Hannukah*. Spring is welcomed in the feast of Passover or *Pesach* marking the early barley harvest, while the Feast of Weeks or Pentecost gives thanks for the wheat harvest. Each of these festivals accrued historical associations and also celebrated

---

1. Scholars such as Mowinckel associate Psalms 96–98 with this feast, due to their allusions to autumn weather.

events in Israel's past, but they began by marking the passing of the seasons and their respective blessings.[2]

The Rhythms of Jesus' Ministry

In John's gospel, the narrative is shaped around the Passover celebrations over a two-year period. There is a year from Jesus in Jerusalem with Nicodemus (2:23) until the feeding of the multitude (6:4).

> The Passover of the Jews was near, and Jesus went up to Jerusalem (2:13).
> When he was in Jerusalem during the Passover festival, many believed in his name because they saw the signs that he was doing (2:23).
> Now the Passover, the festival of the Jews, was near (6:4).

Another year elapses from this event until the meal at Bethany (12:1):

> Now the Passover of the Jews was near, and many went up from the country to Jerusalem before the Passover to purify themselves. (11:55)
> Six days before the Passover Jesus came to Bethany, the home of Lazarus, whom he had raised from the dead. (12:1)

John punctuates his narrative by reference to the annual pilgrim feast of Passover for theological reasons. This, indeed, may be a key John wants us to use to unlock the mystery and meaning of Jesus' death (who, uniquely in John will be put to death when the Passover lambs are killed—Jesus is the new Passover lamb).

During these two years John draws our attention to the passage of the seasons as it is marked by other Jewish festivals. Jesus is set forth as a pilgrim, marking the circling year in the celebration of feasts. There is an unnamed festival in Jerusalem (4:45). Later, he will celebrate the Jewish Feast of Tabernacles (7:2,10). Then came the Feast of Dedication at Jerusalem. John notes: "At that time the festival of the Dedication took place in Jerusalem. It was winter, and Jesus was walking in the Temple, in the portico of Solomon" (John 10:22,23). As Jesus shares in these celebrations, the fourth

---

2. Each of the festivals has acquired a past and future reference. historical remembrances and eschatological, end-time hopes. In spring, Passover recalls the Exodus and longs for ultimate deliverance. Tabernacles, the feast of Booths, recalls the sojourning in tents in Sinai, and looks forward to God enfolding under his canopy all peoples at the end of time. The winter feast of Dedication, Hannukah, recalls the rededication of the Temple and looks forward to the eschatological light of God.

gospel shows us that he was in touch with the rhythm of the seasons and with mother earth yielding her fruits.

Metaphors of Growth: Spiritual Flourishing

In our spiritual journey, we look to creation for descriptors of the inner, spiritual landscape—what Manley Hopkins called the *inscape*, the unique characteristics of the soul. The Hebrew prophets, in particular, employ metaphors facing them each day in their very environment to convey a hopefulness about renewal:

> I will be like the dew to Israel; he shall blossom like the lily, he shall strike root like the forests of Lebanon. His shoots shall spread out; his beauty shall be like the olive tree, and his fragrance like that of Lebanon. They shall again live beneath my shadow, they shall flourish as a garden; they shall blossom like the vine, their fragrance shall be like the wine of Lebanon (Hos 10:12; 14:5–7).

The gospels give us a rich harvest of metaphors of growth, revealing again Christ's deeply sacramental approach to creation. The reign of God germinates as seed in the parables of the sower and the mustard seed; we remember that wheat and tares may grow together (Matt 12, 13). The metaphor about the Vine and branches (John 15:1–11) triggers demanding questions: What is growing well and blossoming, blooming in my spiritual life? Are there buds or potentials that are ready to sprout? Do I notice any dead wood that needs to be taken to the fire and burned? Can I name attitudes or actions that are evidently unproductive and need to be decisively let go of? What parts of my life need pruning or cut back in order to let other parts grow and flourish? What sources of inspiration are my spiritual roots exploring? What is watering the soul? Such an agrarian image invites us to take another look at the root metaphor of growth and fruitfulness, and encourage us to name and celebrate evidences of spiritual maturity and health. Paul delights in the fruit of the Spirit growing in our lives (Gal 5). He says: "You are God's field" (1 Cor 3:9). He develops agricultural imagery: "The point is this: the one who sows sparingly will also reap sparingly, and the one who sows bountifully will also reap bountifully . . . He who supplies seed to the sower and bread for food will supply and multiply your seed for sowing and increase the harvest of your righteousness" (2 Cor 9:6,10).

Seasons of the Spirit

> My beloved speaks and says to me:
> "Arise, my love, my fair one,
>   and come away;
> for now the winter is past,
>   the rain is over and gone.
> The flowers appear on the earth;
>   the time of singing has come." (Song 2:10–12)

The Church has christianized pagan seasonal celebrations, infusing into them fresh meaning. December 25 was the time the Romans marked as the winter solstice, the shortest and darkest day of the year. Jesus was identified with the Sun based on God's ancient promise "for you who revere my name the sun of righteousness shall rise, with healing in its wings" (Mal 4:2). The midwinter feast becomes the celebration of the one referred to in the *Benedictus* canticle:

> By the tender mercy of our God,
>   the dawn from on high will break upon us,
> to give light to those who sit in darkness and in the shadow of death,
>   to guide our feet into the way of peace. (Luke 1:78,79)

The pagan vernal equinox celebration was transformed into the feast of the Annunciation to the Blessed Virgin Mary. The rebirth of spring becomes a celebration of resurrection, the date of Easter being fixed in relation to the new moon. Samhain, the Gaelic festival marking the end of the harvest season and the beginning of winter is transformed into the celebration of the divine light shining out in All Saints.

As we follow the Christian calendar, we glimpse fresh significances in the seasons which speak to us of the climate of the soul. As winter becomes Advent, in the northern hemisphere the earth becoming bare and expectant, so the winter of the soul need not be bleak but a time of deep longing for more of God. We can speak of the springtime of the soul: new shoots of nature point to the possibility of fresh awakenings in the spiritual life. As the sixth century hymn puts it:

> Jesus, the sun of ransomed earth,
> Shed in our inmost souls thy light,
> As in spring days a fairer birth
> Heralds, each morn, the doom of night.

> This hour of grace thou dost impart;
> Teach us with flowing tears the stain
> To cleanse from every victim-heart
> That longs to feel love's welcome pain.
>
> The day is come, the accepted day,
> When grace, like nature, flowers anew;
> Trained by thy hand the surer way
> Rejoice we in our spring-time too.
>
> Let the whole earth in worship bow,
> Great God, before thy mercy-seat,
> As we, renewed by grace, do now
> With praises new thy presence greet.
> (tr. R. A. Knox, 1888–1957)

We experience periods when our prayer can be described as spiritual springtime: budding, flourishing, blossoming—growth and fruitfulness. Themes deriving from the world of nature or agriculture suggest themselves to us now: seeds planted, germinating, fruiting. But at other times our spiritual experience needs to be depicted in terms of barrenness, dryness, as English poet-priest George Herbert asks in *The Flower*:

> Who would have thought my shriveled heart
>   Could have recovered greenness?

The winter of the soul might seem unproductive, but as John of the Cross teaches us, even the dark night of the soul can turn out to be a time of creativity and healing. Poet and mystic Evelyn Underhill (1875–1941) images in her poem *Planting-Time* the soul as a tulip bulb placed deep into the earth, as the soul is hidden with Christ in God (Col 3:3). "Go" she says, "with him/ Into the dim." It is in the dark that God gives the growth, until it is time to bud and blossom forth: the curtained leaves remind Underhill of "the casement of the heart."[3] This evokes the Easter hymn "Now the green blade riseth from the buried grain": seed falling into the ground and dying to enable a great harvest (John 12:24):

> When our hearts are wintry, grieving, or in pain,
> Thy touch can call us back to life again,
> Fields of our hearts that dead and bare have been:
> Love is come again, like wheat that springeth green.
> (J.M.C. Crum, 1872–1958).

---

3. Underhill, *Immanence*, 54.

R.S. Thomas (1913–2000) alludes to a mysterious flower that might represent the experience of prayer:

> The soul
> grew in me
> with its fragrance.
> Men came
> to me from the four
> winds to hear me speak
> of the unseen flower by which
> I sat, whose roots were not
> in the soil, nor its petals the color
> of the wide sea; that was
> its own species with its own
> sky over it, shot
> with the rainbow of your coming and going.[4]

D. H. Lawrence (1885–1930) in his poem *Shadows* hopes for

> snatches of renewal
> odd, wintry flowers upon the withered stem, yet new, strange
>   flowers
> such as my life has not brought forth before, new blossoms of
>   me.[5]

As John of Damascus (d.754) puts it in his Easter song[6]

> All the winter of our sins,
> Long and dark is flying
> From his light . . .
> Now the queen of seasons, bright
> with the day of splendor

Summer is a season of flourishing and richness: new colors and shapes unfold before our eyes, bespeaking spiritual growth. Long, warm days, beckon us to rediscover the sensuality in the natural world, and to seek a sense of fulfillment in the soul.

Autumn, Yeats' "season of mists and mellow fruitfulness," a time of harvest and ingathering, invites us to take stock of the soul. Autumn sums up the paradoxes of the spiritual life. It is both the season of fruitfulness and the beginning of a dying process—echoing the paschal call to let go and let God. Longfellow writes of the *Autumn Within*:

4. Thomas, "The Flower," *Laboratories*.
5. Lawrence, *Complete Poems*.
6. *Come, ye faithful, raise the strain*, tr. John M. Neale (1818–1866).

> It is autumn; not without
> But within me is the cold.
> Youth and spring are all about;
> It is I that have grown old.
> Birds are darting through the air,
> Singing, building without rest;
> Life is stirring everywhere,
> Save within my lonely breast.
> There is silence: the dead leaves
> Fall and rustle and are still;
> Beats no flail upon the sheaves,
> Comes no murmur from the mill.[7]

Yet this may turn out to be the necessary preface to a new beginning…

Dare You Live an Unpredictable Life?

> For everything there is a season, and a time for every matter under heaven:
> a time to be born, and a time to die;
> a time to plant, and a time to pluck up what is planted;
> a time to kill, and a time to heal;
> a time to break down, and a time to build up;
> a time to weep, and a time to laugh;
> a time to mourn, and a time to dance;
> a time to throw away stones, and a time to gather stones together;
> a time to embrace, and a time to refrain from embracing;
> a time to seek, and a time to lose;
> a time to keep, and a time to throw away;
> a time to tear, and a time to sew;
> a time to keep silence, and a time to speak;
> a time to love, and a time to hate;
> a time for war, and a time for peace. (Eccl 3: 1–8)

The spiritual life is marked by times and seasons: the winter of the soul gives way to re-energizing springtime. Temporal metaphors help us celebrate seasons of the spirit: they give us clues with which to read the fecundity or barrenness of the soul. Jesus says: "From the fig tree learn its lesson: as soon as its branch becomes tender and puts forth its leaves, you know that summer is near" (Mark 13:28). But he goes on immediately to

---

7. Longfellow, *Poetical Works*, 783.

warn us against complacency: "Beware, keep alert; for you do not know when the time will come..." (Mark 13:33).

We are increasingly aware that today routines can be disrupted and seasons are showing signs of transition due to climate change. In recent times, the earth's seasons have shifted forwards in the calendar year, with the hottest and coldest days of the years now occurring earlier, and with this change of peak warming and cooling comes a corresponding movement in the onset of the seasons. Scientists have noted other signs that the seasons are shifting: some birds are migrating earlier; plants are blooming sooner; mountain snows are melting quicker. Winters are getting wetter and summers warmer in some places on the planet. Jesus' words take on fresh urgency and relevance both to the climate of the soul and the climate of the planet: "And what I say to you I say to all: Keep awake!" (Mark 13:37).

QUESTIONS FOR REFLECTION

1. What season does your soul find itself in right now? What are the clues, the evidences, the indicators?
2. What signs of growth can you notice in your soul—as you reflect on the last 12 months of your spiritual life?
3. What signs of climate change and shifting seasons are you aware of? What signs of transformation do you notice in the atmosphere of your soul? Is any change inevitable—in the climate exterior or interior? What can be done to impede unwanted change, in either domain?
4. How can you embrace an unpredictable life?

PRAYER EXERCISE

On a fresh piece of paper draw a personal "timeline" to recall the transitions you have personally faced. Draw a horizontal line and mark it into the decades of your life. Above the line, note major events and transitions, including new jobs, house-moves, births and deaths, new ministries. Below the line, try to note how you felt at these moments of change. How did you experience God at these moments? If in a group setting, you might like to reflect on this with a partner. Bring this to a close by giving thanks for God's providence in your life, and entrust your future to him. Conclude by reading aloud Psalm 139:1–18 or use these verses as a prayer of thanksgiving:

"Great is Thy faithfulness," O God my Father,
There is no shadow of turning with Thee;
Thou changest not, Thy compassions, they fail not
As Thou hast been Thou forever wilt be.

"Great is Thy faithfulness!" "Great is Thy faithfulness!"
Morning by morning new mercies I see;
All I have needed Thy hand hath provided—
"Great is Thy faithfulness," Lord, unto me!

Summer and winter, and springtime and harvest,
Sun, moon and stars in their courses above,
Join with all nature in manifold witness
To Thy great faithfulness, mercy and love.

Pardon for sin and a peace that endureth,
Thine own dear presence to cheer and to guide;
Strength for today and bright hope for tomorrow,
Blessings all mine, with ten thousand beside!
(Thomas Obediah Chisholm 1866–1960)

5

Embracing Transition
Winds of Change

You ride on the wings of the wind

(Ps 104:3)

We often experience the disturbing, bracing power of the winds. A walk along the coast or even in the park can be an unsettling, visceral experience as we feel the brisk wind caressing our face and coursing through our hair. Screeching winds tear down and build up. They erode and they deposit, scouring, sculpting and chiseling the landscape. They cleanse the atmosphere of pollutants, dispersing from the air we breathe smoke, dust, and virus, purifying and freshening the atmosphere. Winds scatter seeds and pollen. Jesus referred to the power of the wind: "The rain came down, the streams rose, and the winds blew and beat against that house" (Matt 7:25). As we reflect on the spiritual significance of the wind, we recall phrases that give us clues to the Spirit's work, for we speak of being "free as the wind," we "throw caution to the wind" and we talk of "winds of change." Hildegard sings: "the air is the soul of the earth, moistening it, greening it."[1] Luke gives us a dramatic narrative of his voyages with Paul:

---

1. Uhlein, *Meditations with Hildegard*, 61.

> It was decided that we were to sail for Italy ... Putting out to sea from Sidon, we sailed under the lee of Cyprus, because the winds were against us ... We sailed slowly for a number of days and arrived with difficulty off Cnidus, and as the wind was against us, we sailed under the lee of Crete off Salmone ... sailing was now dangerous ... When a moderate south wind began to blow, they [the crew] thought they could achieve their purpose; so they weighed anchor and began to sail past Crete, close to the shore. But soon a violent wind, called the northeaster, rushed down from Crete. Since the ship was caught and could not be turned with its head to the wind, we gave way to it and were driven. By running under the lee of a small island called Cauda we were scarcely able to get the ship's boat under control. After hoisting it up they took measures to undergird the ship; then, fearing that they would run on the Syrtis, they lowered the sea-anchor and so were driven. We were being pounded by the storm so violently that on the next day they began to throw the cargo overboard, and on the third day with their own hands they threw the ship's tackle overboard. When neither sun nor stars appeared for many days, and no small tempest raged, all hope of our being saved was at last abandoned.... hoisting the foresail to the wind, they made for the beach. But striking a reef, they ran the ship aground; the bow stuck and remained immovable, but the stern was being broken up by the force of the waves. The soldiers' plan was to kill the prisoners, so that none might swim away and escape; but the centurion, wishing to save Paul, kept them from carrying out their plan. He ordered those who could swim to jump overboard first and make for the land, and the rest to follow, some on planks and others on pieces of the ship. And so it was that all were brought safely to land ... After we had reached safety, we then learned that the island was called Malta. The natives showed us unusual kindness. Since it had begun to rain and was cold, they kindled a fire and welcomed all of us round it. (Acts 27)

Luke tells with graphic, frightening detail of the voyage towards Rome. He vividly conveys to his readers both the dangers and the opportunities afforded by the winds. He gives us the impression that the whooshing winds were unpredictable, and they could not stick to a predetermined course. Winds can caress and whisper, or thrash and batter and blast.

Our spiritual experience can accurately be likened to traversing a stormy sea. We need to discern in a time of turbulence whether it is time to hoist the sails, to labor with oars, or to drift along with the wind of the Spirit ... or time to take cover!

Symbols of the Divine

Two contrasting perspectives on the *ruach* or breath of God in the creation accounts are set before us. At the dawn of creation, Genesis tells us, the wind of God blew over the primal chaos of the waters, and human life began when God breathed into the nostrils of man the breath of life. First, the Spirit of God brooded like a mighty wind over the face of the oceans. The second creation account gives us a more tender view, however: "then the Lord God formed man from the dust of the ground, and breathed into his nostrils the breath of life; and the man became a living being" (Gen 2:7).

This anticipates the great passage from the prophets concerning the invigorating and life-giving breath of God. Ezekiel, standing in the valley of dry bones, is commanded:

> Prophesy to the breath, prophesy, mortal, and say to the breath [wind or spirit]: "Thus says the Lord God: Come from the four winds, O breath, and breathe upon these slain, that they may live." I prophesied as he commanded me, and the breath came into them, and they lived, and stood on their feet, a vast multitude. Then he said to me, "Mortal, these bones are the whole house of Israel. They say, 'Our bones are dried up, and our hope is lost; we are cut off completely.' I will put my spirit within you, and you shall live, and I will place you on your own soil; then you shall know that I, the Lord, have spoken and will act, says the Lord." (Ezek 37)

Jesus is Buffeted by the Divine Wind

Mark describes succinctly what happens to Jesus straight after his baptism in the Jordan river: "The Spirit drove him out into the desert, he was with the wild beasts, and the angels ministered to him" (1:12). Of course the Greek word *pneuma* can be translated spirit, breath or wind. The wind of God forced Jesus into the wilderness—a place of radical exposure to the elements and to God. This is a transitional moment for Jesus. The forty days' sojourn echoes vividly the forty years' trek through the desert made by the Israelites in their search for freedom. Now Jesus enters the rocky canyons of the Judean wilderness. Here both hyenas and winds howl! The chiseled and sculpted landscape testifies to the scouring power of the wind in processes of both erosion and deposition. Mirroring the action of the Spirit in the human soul, the wind has a formative influence on the desert, representing how God shapes and reshapes our lives: there are times to build up and times to tear down (Jer 1:10).

Jesus models for us responsiveness to the divine Spirit and reveals how it changed him. The Spirit set his priorities and very direction, for next "Jesus returned in the power of the Spirit into Galilee" (Luke 4:14). There, in Nazareth, Jesus quotes Isaiah 61 to reveal the source of his inspiration: "The Spirit of the Lord is upon me . . . to proclaim release to the captives." Throughout his ministry, the wind of the Spirit will be his driving force and leading him into joy: "in that same hour he rejoiced in the Holy Spirit" (Luke 10:21).

When Luke describes the drama of Pentecost he relates: "When the day of Pentecost had come, they were all together in one place. And suddenly from heaven there came a sound like the rush of a violent wind, and it filled the entire house where they were sitting" (Acts 2:1,2). But which of the winds of the Holy Land does he have in mind? Is this a calming gentle breeze or a gusting, stormy gale? What type of wind buffeting Jerusalem does Luke have in view? The different winds illustrate the different workings of the Spirit in our life.

THE FOUR WINDS OF THE HOLY LAND

The Bible talks about the four winds. Sometimes they represent the universality of God and his sovereignty over all the earth, over every point of the compass. Sometimes they represent differing perspectives on the Divine. Jesus teaches: "he will send out the angels, and gather his elect from the four winds, from the ends of the earth to the ends of heaven" (Mark 13:27). John relates his vision: "After this I saw four angels standing at the four corners of the earth, holding back the four winds of the earth so that no wind could blow on earth or sea or against any tree" (Rev 7:1). The tradition of the four winds emerges in the Hebrew scriptures. We read: "I, Daniel, saw in my vision by night the four winds of heaven stirring up the great sea" (Dan 7:2). Jeremiah sees them as sign of cleansing judgment: "I will bring upon Elam the four winds from the four quarters of heaven; and I will scatter them to all these winds, and there shall be no nation to which the exiles from Elam shall not come" (Jer 49:36). Zechariah depicts the winds as being at God's service: "The angel answered me, 'These are the four winds of heaven going out, after presenting themselves before the Lord of all the earth'" (Zech 2:6, 6:5).

From the North

The wind from the north is dry and cold. Job (37:9) tells us:

> From its chamber comes the whirlwind,
> and cold from the scattering winds.

Eccesiasticus describes its effects: "The cold north wind blows, and ice freezes on the water; it settles on every pool of water, and the water puts it on like a breastplate" (Sirach 43:20). We should not be surprised at the bitter cold of the north wind, for Jeremiah asks us: "Does the snow of Lebanon leave the crags of Sirion? Do the mountain waters run dry, the cold flowing streams?" (18:14). Is this the wind Luke has in mind when talking of the experience of the Spirit?

From the South and East

The dry scorching wind unsettles and desiccates all in its path, bearing a mist of erosive fine sand. This is the *sirocco*, originating over the deserts of northern Africa and Arabia, blotting out the sun and scorching vegetation. A Victorian traveler relates: "the glow of the wind came upon our faces as from a burning oven."[2] Other writers testify to this wind as "a blast from a furnace." Jesus knows this well: "And when you see the south wind blowing, you say, 'There will be scorching heat'; and it happens" (Luke 12:55). In the prophets, this blistering wind becomes a symbol of judgment:

> At that time it will be said to this people and to Jerusalem: A hot wind comes from me out of the bare heights in the desert towards my poor people, not to winnow or cleanse— a wind too strong for that (Jer 4:11,12; cf. Nah 1:2–8).
> When it is transplanted, will it thrive?
> When the east wind strikes it,
>   will it not utterly wither,
>   wither on the bed where it grew? (Ezek 17:10)
> But it [the vine of Israel] was plucked up in fury,
>   cast down to the ground;
> the east wind dried it up;
>   its fruit was stripped off,
> its strong stem was withered;
>   the fire consumed it. (Ezek 19:12)
> Although he may flourish among rushes,
>   the east wind shall come, a blast from the LORD,
>   rising from the wilderness;
> and his fountain shall dry up,
>   his spring shall be parched.
> It shall strip his treasury
>   of every precious thing. (Hos 13:15)

---

2. Robinson, *Biblical Researches*, 287.

The book of Genesis talks of the heads of grain that sprouted and then got withered by the scorching east wind (41:6, 41:23, 41:27). The east wind of the Bible is a fierce wind (Isa 27:8, Job 38:24), destroying ships on the high seas (Ps 48:7, Ezek 27:26), and scattering people (Job 15:2, 27:21, Jonah 4:8, Jer 18:17).

This wind speaks of the Holy Spirit the discomforter! The Holy Spirit can be a disturbing and unsettling presence in our lives. If we find ourselves complacent, settled, or stuck in our praying, then we need such a Spirit and such a wind to move us on, somehow, in our spiritual life. Interestingly, the dry, sand-bearing easterly wind is called the *hamsin*, from the Arabic word for "fifty," because it blows especially over the fifty-day period from Easter to Pentecost! It is the actual wind of "Pentecost"—which itself comes from the Greek word in the Septuagint meaning "fiftieth"—referring to the festival celebrated on the fiftieth day after Passover, so *hamsin* translates "Pentecost." There is special symbolism here. The *sirocco* or *hamsin* wind makes people vulnerable, creating an acute thirst for replenishing water, and has the potential to reshape and resculpture the very landscape—of the soul!

From the West

Picking up moisture as it blows across the Mediterranean, the westerly wind brings welcome refreshment, and often comes to Jerusalem in the late afternoon, cooling its residents after a hot day. In winter months it deposits its moisture on the Mount of Olives, the watershed: this cascades down into the Judean wilderness to the east. This brings renewal to those struggling in the blazing desert heat: bespeaking the Holy Spirit the Comforter, the one who restores the weary. For most of the year the desert is a parched, baked landscape where grasses turn brown, burnt in the sun. Then, unexpectedly come the winter winds and their generous gift of rain. When there is a downpour on the Mount of Olives, the water cascades eastwards towards the great rift valley and flash floods sweep through the desert ravines, shifting boulders and cutting the edges of the valley. When the wet wind lashes the desert it is very welcome indeed: it is the harbinger of spring, turning back the drought and enabling new shoots and fresh growth of plants, the miraculous spreading out of a green mantle across the desert. This speaks to us of the Holy Spirit who comes to re-energize us, awakening us into new life.

Winds of Change

Dare we pray with the Song of Songs (4:16)

> Awake, O north wind,
>   and come, O south wind!
> Blow upon my garden
>   that its fragrance may be wafted abroad.

We may not predict the moving of the Wind in our spiritual lives but we must expose ourselves to its presence: "the wind blows where it chooses, and you hear the sound of it, but you do not know where it comes from or where it goes. So it is with everyone who is born of the Spirit" (John 3:8). Four themes emerge in this key saying of Jesus:

- **Mystery**: the Spirit, like the wind is untamable and unpredictable. You can't box in the Divine or tie God down. Like wind the Spirit is invisible, though you certainly feel his effects.[3]

- **Movement**: the Spirit blows in our lives, if we let him, to create movement in the soul—to shift us, to change us. We are invited to move from resistance to surrender, from bitterness to forgiveness, from anxiety to trust, and from anger to acceptance, from egocentricity to selflessness. We are invited to welcome at the center of our lives the transitions and shifts that will make us more Christlike.

- **Momentum**: the wind of the Spirit, as the very breath of God, is lifegiving. He is energy for the soul. Jesus promised: "You shall receive power when the Spirit comes" (Acts 1:8)—the word for "power" is *dunamis* from which we get the term "dynamite"! This promise was fulfilled when the mighty wind blew at Pentecost, transforming timid and fearful disciples into apostles that would turn the world upside down as they went out to proclaim the Good News (Acts 17:6).

- **Molding**: "you hear the sound of it." We hear the wind whistling through the trees, howling across the waters, rustling and whispering in the leaves. In the Judean desert you can see how wind has both carved crevices out the soft rock and also deposited particles in new creations. God's Spirit longs to shape and reshape our lives. Like a landscape open to the wind we are invited to bare our soul to God who can do wonderful things with the "raw material" of a human life yielded to his hands. Spiritual formation is a process by which a person gets reshaped: the metaphor of formation is drawn from the natural world, speaking of a creative process at work in the landscape both physical and spiritual.[4] It implies that at the heart of spirituality is the

---

3. The Spirit can be imaged as feminine, too, especially as the Hebrew *ruach* is feminine; the Greek *pneuma* is not masculine but neuter.

4. The language of formation is a hylomorphic term, utilized scientifically of the

raw material of a person's life, on which God acts in a creative way. Robert Mulholland puts it succinctly: "Spiritual formation is the experience of being shaped by God towards wholeness."[5] As we reflect on this, we realize that wind encounters unyielding materials and rocks in its path. What resistances are we putting up before God?

God literally inspires us—for the word "inspire" from the Latin *inspirare* means "blow into, breathe upon." In the act of new creation, Jesus reproduces this act:

> Jesus said to them again, "Peace be with you. As the Father has sent me, so I send you." When he had said this, he breathed on them and said to them, "Receive the Holy Spirit." (John 20:21,22)

In the upper room the risen Lord animates and re-energizes his disciples.

SPIRITUAL WRITERS DISCOVER THE DIVINE WINDS

Macarius of Egypt in the fourth century delighted in metaphors for the Spirit as he sought to describe the experience of encountering deeply the Spirit. To him are attributed an image of two conflicting winds:

> As a certain strong wind blows in a dark and gloomy night and strikes all the plants and seeds, moving them this way and that in shaking agitation, so also man, who has fallen under the power of the night of the devil of darkness. He also lies in night and darkness, is moved, buffeted and is shaken by the stiff blowing wind of sin and through all his nature, namely, in his soul, thoughts and mind, he is thoroughly affected. All the members of the body are shaken, not one part of the soul or the body is immune from the passions of sin dwelling in us.
>
> In a similar way there is a day of light and the divine wind of the Holy Spirit, breathing through and refreshing souls who live in the day of the divine light. It passes through the whole nature of the soul, the thoughts and the entire substance of the soul and all the members of the body, as it recreates and refreshes them with a divine and ineffable tranquility . . .[6]

---

shaping of matter. Used of spiritual remolding it is an analogy with the shaping of matter. See Kelsey, "Reflections."

5. Mulholland, *Invitation*.

6. Maloney, *Intoxicated with God*, 35. Scholars often attribute the homilies to a "pseudo Macarius" as the identity of the author is uncertain.

Hildegard of Bingen testifies to cosmological visions alert and responsive to the physical winds coursing across the planet. In her *Book of Divine Works* she shares her understanding:

> As you see, the east wind and the south wind, together with their sidewinds, stir up the firmament with mighty gusts and let themselves be blown across the Earth from east to west. This has the following meaning: when the breath of the fear of the Lord and the breath of God's judgment affect our inner mind through the other powers of virtue in the might of holiness, they give rise to our mind in the fair East and build it up to a fair completion as far as the fair West. They allow us to remain victors over the things of the flesh and in this way to persevere . . . and the same thing occurs with respect to the other winds: just as the wind stirs the humors, fear of God stirs the human conscience—it is a warning to us to stride ahead on the path of virtue. In this connection we often run into bodily stress and become weary of good works. But then, along with the south wind, God's grace comes close to us in a gracious way and guides our spirit in the earthly struggle.[7]

Following ancient medical tradition, Hildegard understands human well-being in terms of four bodily fluids or humors, physically affected by the winds. But this becomes a metaphor for the flourishing of inner temperaments: she discerns in the movement of the winds clues pointing to the movements of the soul. Our openness to "stirrings" of the Spirit, and our readiness to welcome the breezes of God's grace is, for Hildegard, the very key to discovering healing and wholeness.

Edwin Hatch celebrates this metaphor in his great hymn (1848):

> Breathe on me, breath of God,
> Fill me with life anew,
> That I may love what Thou dost love,
> And do what Thou wouldst do.

Gerard Manley Hopkins develops the metaphor in daring ways in his poem "The Blessed Virgin compared to the Air we Breathe." He calls her

> Wild air, world-mothering air,
> Nestling me everywhere.

Wind—and air—become powerful images of God's providence and mysterious presence.

---

7. Fox, *Divine Works*, 63.

Celtic Christians experienced full-on an exposure to the elements, with their venturesome traditions of *peregrination* and voyaging on the rough seas around Ireland and Scotland. They were motivated by a desire both to spread the gospel and to discover God's providence in the deep, as depicted in the sixth century *Voyage of Brendan*. The Celtic tradition brings together the elemental metaphors speaking of divine and human:

> *God*
> I am the wind that breathes upon the sea,
> I am the wave on the ocean,
> I am the murmur of leaves rustling,
> I am the rays of the sun . . .
> *The soul*
> I am a flame of fire, blazing with passionate love;
> I am a spark of light, illuminating the deepest truth . . .
> I am a wild storm, raging at human sins;
> I am a gentle breeze, blowing hope in the saddened heart . . . [8]

Harnessing the Power of the Wind

From ancient times, the power of the wind for good has been recognized in the Holy Land. Ruth is found winnowing barley at the threshing floor of Boaz (Ruth 3). The very site of the Temple was the threshing ground of Araunah the Jebusite according to 2 Sam 24 and 1 Chr 21, purchased by David. Its location on the exposed hilltop of Mount Moriah (later designated Mount Zion) ensured that it functioned well as a place to catch the wind in order to separate the wheat from the chaff.

In ancient times, the great sails of sea-going ships harnessed the power of the wind. King Solomon built a fleet of ships at Ezion-geber, on the shore of the Red Sea (1 Kgs 9), and his ships imported precious metals from the port of Tarshish on the Mediterranean and cedar of Lebanon for the Temple from the Phoenician seaports of Tyre and Sidon. Psalm 107 celebrates:

> Some went down to the sea in ships,
>   doing business on the mighty waters;
> they saw the deeds of the LORD,
>   his wondrous works in the deep.
> For he commanded and raised the stormy wind,
>   which lifted up the waves of the sea. (Ps 107:23–25)

---

8. "The Black Book of Carmarthen," de Weyer, *Celtic Fire*, 92.

In modern times, the state of Israel has built wind-farms on the Golan Heights and on Mount Gilboa, mighty turbines catching the power of the wind and generating significant amounts of power.

How are we to catch the wind of God's Spirit? How can we receive the energizing grace of the breath of God? It comes back to the theme of exposure. We must locate ourselves, like a threshing floor or sails of a boat, in such a way as to be orientated to the divine Wind and Breath.

QUESTIONS FOR REFLECTION

1. What is the climate of your soul right now? Do you need
   - a cleansing wind, like that from the north, that will blow away the cobwebs, the debris and detritus and clutter in your life?
   - a challenging wind, like the sirocco, that will unsettle and disturb complacency?
   - a refreshing, renewing wind, like the westerly in the Holy Land, that will restore, invigorate, re-energize?

   How would you express your need for the Spirit?

2. "The wind blows where it chooses, and you hear the sound of it, but you do not know where it comes from or where it goes. So it is with everyone who is born of the Spirit" (John 3:8). How do you find yourself responding to Jesus' words about the unpredictable Spirit?

3. Hildegard of Bingen described herself as "a feather of the breath of God." What phrase or metaphor might convey your own sense of identity before God and his breezes?

4. If a key aspect of the wind concerns movement, what shifts can you name in the course of your spiritual life in the last year? In what ways are you becoming more open to the winds of God?

5. In the twelfth century Hildegard wrote: "The winds are burdened by the utterly awful stink of evil, selfish goings-on . . . The air belches out the filthy uncleanliness of the people. There pours forth an unnatural, loathsome darkness that withers the green, and wizens the fruit that was to serve as food for the people."[9] What is polluting today's moral atmosphere, and how do you find yourself responding to this?

---

9. Uhlein, *Meditations with Hildegard*, 10.

PRAYER EXERCISE

Take a walk outside and experience the wind's moods. Allow the breeze to caress you or disturb you. Open your soul to the breath of God. Use your breathing as a prayer, both your exhaling and inhaling. Breathe out all negativity and anxiety. Breathe in God's Spirit, his very breath. You can employ the rhythms of breathing using the Jesus Prayer. The first part "Lord Jesus Christ, Son of God" is said while drawing in the breath and the second part "have mercy on me a sinner" is synchronized with exhalation. As you use such breath prayers, recall that prayer is the oxygen of the soul![10] Conclude with a prayer to Holy Spirit using a classic or modern hymn:

The words by H.W. Baker (1821–1877):

> Life-giving Spirit, o'er us move, as on the formless deep;
> Give life and order, light and love,
> Where now is death or sleep ...
> True wind of heaven, from south or north,
> For joy or chastening blow;
> The garden-spices shall spring forth
> If thou wilt bid them flow.

Or Jane and Betsy Clowe's song:

> Wind, wind, blow on me,
> Wind, wind, set me free,
> Wind, wind,
> The Father sent the blessed Holy Spirit.
> When we're weary You console us;
> When we're lonely You enfold us;
> When in danger You uphold us,
> Blessed Holy Spirit.

Or compose your own prayer, inspired by Manley Hopkins or by the hymn writers.

---

10. Gillet, *Jesus Prayer*; French, *Way of a Pilgrim*.

6

Braving Exposure
The Summons of the Sun

The spirit of the Lord speaks through me,
    his word is upon my tongue . . .
One who rules over people justly,
    ruling in the fear of God,
is like the light of morning,
    like the sun rising on a cloudless morning,
    gleaming from the rain on the grassy land.

(2 Sam 23:2–4)

"I give you thanks, for to me you are a light that knows no evening, a sun that never sets." So prays the great mystical writer of the eastern church, Symeon the New Theologian (949–1022). Blazing, dazzling, flaming, glowing, blistering, shimmering, scorching—the radiant image of the sun is a natural metaphor for the Divine. In this chapter we also explore its potential to describe the soul that is fully alive, and see how it can become an image for our very vocation. We consider how far we are prepared to risk exposure to the Divine! We recall that traditionally churches, through the centuries, were built to face east—at least in the northern hemisphere—precisely to welcome the dawn and to encourage worshippers to orientate themselves to the sun!

The Wisdom text Ecclesiasticus celebrates the splendor of the sun:

> The pride of the higher realms is the clear vault of the sky,
> > as glorious to behold as the sight of the heavens.
>
> The sun, when it appears, proclaims as it rises
> > what a marvelous instrument it is, the work of the Most High.
>
> At noon it parches the land,
> > and who can withstand its burning heat?
>
> A man tending a furnace works in burning heat,
> > but three times as hot is the sun scorching the mountains;
>
> it breathes out fiery vapors,
> > and its bright rays blind the eyes.
>
> Great is the Lord who made it;
> > at his orders it hurries on its course. (Sir 43:1–5)

THREE THEMES IN THE SCRIPTURES

1 Light and Life

> God said, "Let there be light"; and there was light. And God saw that the light was good; and God separated the light from the darkness. God called the light Day, and the darkness he called Night. And there was evening and there was morning, the first day.... And God said, "Let there be lights in the dome of the sky to separate the day from the night; and let them be for signs and for seasons and for days and years, and let them be lights in the dome of the sky to give light upon the earth." And it was so. God made the two great lights—the greater light to rule the day and the lesser light to rule the night—and the stars. God set them in the dome of the sky to give light upon the earth, to rule over the day and over the night, and to separate the light from the darkness. And God saw that it was good. (Gen 1:3–5,14–18)

Light from the sun is the primordial gift of life on earth. As the Psalmist celebrates:

> In the heavens he has set a tent for the sun,
> > which comes out like a bridegroom from his wedding canopy,
> > > and like a strong man runs its course with joy.
> >
> > Its rising is from the end of the heavens,
> > > and its circuit to the end of them;
> >
> > and nothing is hidden from its heat. (Ps 19: 4–6)

God himself is hailed as the sun: "For the Lord God is a sun and shield; he bestows favor and honor" (Ps 84:11). The prophet Isaiah highlights a sense of clarity and purpose in the gift of sunlight: "For thus the Lord said to me: I will quietly look from my dwelling like clear heat in sunshine, like a cloud of dew in the heat of harvest" (18:4). Later, he will describe the healing power of God's light: "Moreover, the light of the moon will be like the light of the sun, and the light of the sun will be sevenfold, like the light of seven days, on the day when the Lord binds up the injuries of his people, and heals the wounds inflicted by his blow" (Isa 30:26). This is a theme echoed by Malachi: "But for you who revere my name the sun of righteousness shall rise, with healing in its wings" (Mal 4:2). In the NT, Jesus' birth is indicated by a shining star in the inky blackness of the night sky. Zechariah, father of John the Baptist had proclaimed:

> By the tender mercy of our God,
>   the dawn from on high will break upon us,
> to give light to those who sit in darkness and in the shadow of death,
>   to guide our feet into the way of peace. (Luke 1:78,79)

Peter echoes this hope when he writes of Christ's return: "the day dawns and the morning star rises in your hearts" (2 Pet 1:19).

2 Glory and Hope

Dramatically, the prophet Habakkuk announces the advent of God:

> God came from Teman,
>   the Holy One from Mount Paran.
> His glory covered the heavens,
>   and the earth was full of his praise.
> The brightness was like the sun;
>   rays came forth from his hand,
>   where his power lay hidden. (3:3–5)

The sun is a natural image with which to convey the wonder and mystery of divine theophany. Matthew describes the transfiguration of Christ in these terms: "And he was transfigured before them, and his face shone like the sun, and his clothes became dazzling white" (17:2). This is echoed in the vision of John in the book of Revelation:

> In his right hand he held seven stars, and from his mouth came
> a sharp, two-edged sword, and his face was like the sun shining

with full force. When I saw him, I fell at his feet as though dead. But he placed his right hand on me, saying, "Do not be afraid; I am the first and the last, and the living one. I was dead, and see, I am alive for ever and ever; and I have the keys of Death and of Hades" (1:17,18).

3 Dangers and Risks

Those who live in the Middle East are acutely aware of the dangerous aspects of exposure to the searing heat of the sun. This features in Jonah's tale: "When the sun rose, God prepared a sultry east wind, and the sun beat down on the head of Jonah so that he was faint and asked that he might die. He said, 'It is better for me to die than to live'" (Jonah 4:8). In his parables, we noted, Jesus twice refers to the burning heat of the sun. Describing the seed scattered among rocks Jesus says: "And when the sun rose, it was scorched; and since it had no root, it withered away" (Mark 4:6). In the parable of the workers in the vineyard, the day laborers complain about the reward given to late-comers: "These last worked only one hour, and you have made them equal to us who have borne the burden of the day and the scorching heat" (Matt 20:12). James (1:11) takes a metaphorical approach: "For the sun rises with its scorching heat and withers the field; its flower falls, and its beauty perishes. It is the same with the rich; in the midst of a busy life, they will wither away."

Temperature of the Soul

The biblical writers know from first-hand experience that while clear skies allow the blaze of sunlight, they also facilitate the biting cold at night. Jacob relates the physical reality: "It was like this with me: by day the heat consumed me, and the cold by night, and my sleep fled from my eyes" (Gen 31:40). Job brings a shiver to our spine: "They lie all night naked, without clothing, and have no covering in the cold" (Job 24:7) while the Psalmist asks: "He hurls down hail like crumbs— who can stand before his cold?" (Ps 147:17).

This becomes a powerful metaphor for the temperature of the soul. Jesus declares: "because of the increase of lawlessness, the love of many will grow cold" (Matt 24:12). In Revelation Christ penetratingly says: "I know your works; you are neither cold nor hot. I wish that you were either cold or hot. So, because you are lukewarm, and neither cold nor hot, I am about to spit you out of my mouth" (Rev 3:15,16).

Such language runs through John's gospel. The darkness cloaking Nicodemus' quest is symbolic, denoting the state of his soul: "he came to Jesus by night" (John 3:2). In the passion accounts, the evangelist develops this motif. On Maundy Thursday, when Judas gets up from the table of the Last Supper and goes out to betray Jesus to the authorities, John notes: "So, after receiving the piece of bread, he immediately went out. And it was night" (13:30). This is not simply a time reference—it conveys the blackness of Judas' soul, and the darkness of his intentions. Jesus had earlier said: "The light is with you for a little longer. Walk while you have the light, so that the darkness may not overtake you. If you walk in the darkness, you do not know where you are going. While you have the light, believe in the light, so that you may become children of light" (12:35–36). But now, with Judas' betrayal, it is pitch-black darkness.

John goes on to talk about the physical temperature: "Now the servants and the police had made a charcoal fire because it was cold, and they were standing round it and warming themselves. Peter also was standing with them and warming himself" (18:18). John repeats this detail: "Then Annas sent him bound to Caiaphas the high priest. Now Simon Peter was standing and warming himself" (18:24,25). John sets up a vivid contrast, and the physical conditions denote spiritual realities. Peter is enclosed in the courtyard of the high priest, while Jesus is examined by Annas. Peter is comforting himself by the warmth of the fire while Jesus is led outside in the biting cold of the night to Caiaphas. Peter experiences a chill and shiver in his soul and finds relief in the enclosure, but Jesus, in his vulnerability faces exposure. One is turned in on himself in selfish preoccupation and self-preservation; the other goes out to lay down his life on the Cross.

When we turn to the other gospels we see that they employ allusions to the weather at the crucifixion, to evoke the meaning of what is going on. Luke says: "It was now about noon, and darkness came over the whole land until three in the afternoon, while the sun's light failed" (23:44,45). This can be translated "darkness descended over the whole earth . . . the sun was eclipsed." Matthew adds a further detail: "Then Jesus cried again with a loud voice and breathed his last. At that moment . . . the earth shook, and the rocks were split . . . Now when the centurion and those with him, who were keeping watch over Jesus, saw the earthquake and what took place, they were terrified and said, 'Truly this man was God's Son!'" (Matt 27:50,51,54).

The evangelists are talking about signs in the heavens and signs on the earth as they invite us to take our place at the foot of the Cross. What does this mean—the sun's eclipse, the earthquake? It certainly means that what is happening is not the local crucifixion and punishment of a criminal or preacher. What is happening is a cosmic, seismic event—it has importance for the whole universe—it affects the very heavens and the earth. It means—God is doing something dramatic and significant. Our response should be holy fear, a sense of being awestruck—we are allowed to become dumbfounded, lost for words as we stand before the Cross of Christ.

We next encounter chilly darkness in John's gospel on Easter morning: "Early on the first day of the week, while it was still dark, Mary Magdalene came to the tomb" (20:1). The darkness denotes Mary's sense of grief and confusion—after all, Jesus had said: "If you walk in the darkness, you do not know where you are going." Mary was walking in a daze, heartbroken and desolate in her bereavement. The other gospels bring a glimmer of light: "And very early on the first day of the week, when the sun had risen, they went to the tomb" (Mark 16:1). "The sun had risen"—the pun is in English, not in the Greek. But a new day is dawning—indeed a new future of the world is happening in Easter daybreak.

Exposure: Jesus and the Desert Fathers and Mothers

Jesus encounters the blistering, searing power of the sun and its message when he is driven to the desert by the Spirit after his baptism in the river Jordan. As we recalled, Mark puts it succinctly: "he was with the wild beasts, and the angels ministered to him" (1:12). For Jesus, the desert turns out to be a place of wild beasts where he encounters jackal and hyena. There are scorpions underfoot. He is exposed to the elements: to howling wind and unforgiving, blazing sun. But he is exposed to other dangers, too: he must grapple with powerful temptations, and struggle with demons. Peter puts it: "Your adversary the devil prowls around like a roaring lion, looking for someone to devour" (2 Pet 5:8). As the parable of the Good Samaritan reminds us, the desert is also the place of bandits and terrorists. It is a risky place in every sense. And yet Jesus is also enfolded in the care of angels and discerns the Father's voice.

The desert landscape spoke to him of what is real in the human condition. The burning bright intensity of the sun shows up places of light and shade, and the shade is so tempting. Jesus saw vividly and dramatically the choices open to everyone: the risk of staying out in the sun, and the seductive shadows. "Although he was a Son, he learned obedience through what he suffered" (Heb 5:8). His sojourn in the desert shows Jesus' radical solidarity with us in our desert places.

The desert becomes a powerful symbol of Christian prayer, because it is a place of truth, of radical, searing honesty. There is no place for pretense, for role-playing, for the wearing of masks before God. The sun burns searingly into the soul. In the desert as we step out into the open, we are exposed to elements human and physical. This is a place of spiritual nakedness. Jerome said: "the desert strips you bare." In the desert of the spirit, too, prayer gets real. Our self-protective barriers and defenses must crumble before God. The false, competitive self must die. The self or ego identified with our *persona* (Greek—mask) that we present to the world must wither and fade away. The image of ourselves that we like others to see—confident, competent—is often shaped or conditioned by our culture, by advertisements, by the modern preoccupation with image or cult of celebrity, by the compulsions and illusions of our age. People like a good performance. People want to see beautiful bodies, well dressed, unwrinkled, attractive. People admire the ones who seem to be wealthy and successful. They worship images of perfection. This is a self-image or image of the self that we would like to project to others. We think that our worth, our value, comes from what other people say about us, how they acclaim us and appreciate us. But this is the illusory self, clamoring for attention. It is a wax mask that must melt in the heat of the desert.

The desert fathers and mothers went into the desert to discover God: in the process they discovered themselves too. Seekers came to request a word of counsel and advice saying "Give me a word, Father." Often the word offered penetrates all pretenses and role-playing, slices through masks and artificialities, and cuts to the quick, exposing falsities. In the desert the fathers recognize demons and angels. In the dazzling light of the day, they are ready to expose human foibles and self-importances:

> One day when Abba John was sitting in front of the church, the brethren were consulting him about their thoughts. One of the old men who saw it became a prey to jealousy and said to him, "John, your vessel is full of poison." Abba John said to him, "That is very true, abba; and you have said that when you only see the outside, but if you were able to see the inside, too, what would you say then?"[1]

Here the passions and lusts, fantasies and temptations are magnified and seen in all their ferocity as undermining true identity in God. Anthony the Great said: "He who wishes to live in solitude in the desert is delivered from

---

1. Ward, *Sayings*, 87.

three conflicts: hearing, speech, and sight; there is only one conflict for him and that is with fornication."[2]

We must come to terms with our weaknesses. Maybe the angels and demons are within us: not external entities but interior aspects of our soul. The desert fathers sometimes over-estimated their capabilities:

> Abba John said to his older brother: "I should like to be free of all care, like the angels, who do not work but ceaselessly offer worship to God." So he took off his cloak and went away into the desert. After a week, he came back to his brother. When he knocked on the door, he heard his brother say, before he opened it, "Who are you?" He said, "I am John, your brother." But he replied, "John has become an angel, and henceforth he is no longer among men."

Finally, opening the door to him the next day, the elder brother said to him, "You are a man and must once again work in order to eat." Then John made a prostration before him, saying, "Forgive me."[3]

The desert of prayer summons us towards increasing exposure, to the reality of sin and temptation, to the reality of the self, and above all exposure to God's blazing presence. In a memorable episode from the Desert Fathers we learn the soul's true potential:

> Abba Lot went to see Abba Joseph and said to him, "Abba, as far as I can I say my little office, I fast a little, I pray and meditate, I live in peace and as far as I can, I purify my thoughts. What else can I do?" Then the old man stood up and stretched his hands towards heaven. His fingers became like ten lamps of fire and he said to him, "If you will, you can become all flame."[4]

Our potential and vocation is to be ignited by the Spirit, engulfed with his fire, radiant and ablaze with the Spirit himself, the divine flame. What is your experience of the divine fire?

Early Christian Hymns of Light

The first generations of Christians delighted to sing of Christ the light. *Hail Gladdening Light* is a third century hymn sung at the lighting of lamps at dusk:

---

2. Ward, *Sayings*, 3.
3. Ward, *Sayings*, 86.
4. Ward, *Sayings*, 103.

> Hail, gladdening Light, of His pure glory poured
>   Who is the immortal Father, heavenly, blest,
>   Holiest of Holies, Jesus Christ our Lord!
>
> Now we are come to the sun's hour of rest;
>   The lights of evening round us shine;
>   We hymn the Father, Son, and Holy Spirit divine!
>
> Worthiest art Thou at all times to be sung
>   With undefiled tongue,
>   Son of our God, Giver of life, alone:
>   Therefore in all the world Thy glories, Lord, they own
> (tr. John Keble)

In the next century Ephrem the Syrian hymnwriter associates Jesus with *Sol Invictus*—the Unconquered Sun—the sun-god of the Roman empire:

> Light was like a harbinger
> To that Bright One to Whom Mary gave birth,
> For His conception was in the victory of light,
> And His birth was at the victory of the sun.
> Blessed be the Conqueror![5]

Day and Night in Christian Spirituality

> Abide with me: fast falls the eventide;
> the darkness deepens; Lord, with me abide:
> when other helpers fail and comforts flee,
> help of the helpless, O abide with me.

These words by Henry Francis Lyte (1847) reverberate strongly with those going through the pain of bereavement. The most fundamental metaphors of time, of course, relate to night and day. William Blake had written menacingly in *The Tyger* of "the forests of the night"[6] and Rainer Maria Rilke alludes to the ambiguity of darkness in *Du Dunkelheit*. Night is not necessarily symbolic of death and gloom; it can become positive, since God has a habit of working in the dark!

The eastern tradition of spirituality celebrates the uncreated light of Tabor—the *Metamorphosis*. Lossky comments: "To see the divine light with

---

5. McVey, *Ephrem*, 210.
6. Erdman, *Complete Poetry*.

bodily sight, as the disciples saw it on Mount Tabor, we must participate in and be transformed by it, according to our capacity. Mystical experience implies this change in our nature, its transformation by grace."[7] Dare we enter the divine light—even participate in the energies of God—if it might alter us, reshape us, make us different? In the eleventh century St Symeon the New Theologian offered this prayer to the Holy Spirit:

> You are the exaltation, and You are the merriment, my God;
> Your grace, the grace of the All-Holy Spirit
> shall shine like the sun in all the saints.
> And You, the unapproachable Sun
> shall shine in the middle of them.[8]

Times of transition especially resonate with the shifting states of day. Twilight and dusk, with lengthening shadows and the descent of darkness, may evoke depression or disorientation. With the Psalmist we find ourselves yearning for a new dawn: "my soul waits for the Lord more than those who watch for the morning" (130:6). Charles Wesley in his hymn of 1740 powerfully expresses this theme:

> Christ, whose glory fills the skies, Christ the true, the only Light,
> Sun of Righteousness, arise! Triumph o'er the shades of night:
> Dayspring from on high, be near; Daystar, in my heart appear.
>
> Dark and cheerless is the morn unaccompanied by thee;
> joyless is the day's return, till thy mercy's beams I see,
> till they inward light impart, glad my eyes, and warm my heart.
>
> Visit then this soul of mine! Pierce the gloom of sin and grief!
> Fill me, Radiancy Divine; scatter all my unbelief;
> more and more thyself display, shining to the perfect day.

Hildegard sees the sun both as an image of the Divine and as a symbol of the human vocation to peruse justice. She sings:

> The sun is set in the firmament of heaven.
> It watches over earthly creation,
> letting nothing perish.
> God watches over us in just such a way.
> In no way,
> can a believer that sets
> heart and being on God

---

7. Lossky, *Mystical Theology*, 223.
8. Symeon, "Hymn 1," *Divine Eros*.

> ever be forgotten by God.
> The sun climbs in its course
> and at mid-day burns in full glow.
> This is how it is with those who
> are just.
> They demonstrate fullness and justice.
> They cannot be hindered,
> just as the sun cannot be hindered
> in its ascent.
> No warmth ever goes to waste.
> The earth is to the sun,
> as the soul is to God.[9]

Mechthild of Magdeburg communicates in *The Flowing Light of the Godhead* a vision that is cosmic as well as intensely personal:

> The Godhead is so blazing hot,
> As you well know,
> That all the fire and all the glowing embers
> That make the heavens and all the saints glow and burn
> Have flowed out from his divine breath . . .
> He can both burn powerfully and cool consolingly.

Mechthild writes of the climate of the soul:

> God shines beautifully into all according to the degree of holiness they achieved here in love and according to how noble in virtues they become . . . the sun shines according to the weather. There are different kinds of weather on earth under the sun . . . Hence, he is to me as I can bear him and see him.[10]

Facing the Shadows

> I go about in sunless gloom;
> I stand up in the assembly and cry for help.
> Have you . . . walked where the Abyss is deepest?
> Have you been shown the gates of Death
> or met the janitors of Shadowland? (Job 30:28; 38:16,17, JB)

As Jung and others explored this idea, they saw that shadows may not necessarily be entirely negative. Sure, they can represent those darker sides of

---

9. Uhlein, *Meditations with Hildegard*, 53,54.
10. Tobin, *Mechthild*, 154.

our personality that we do not wish to show to the world, so we suppress them and alienate them from our day-to-day operations and consciousness: things like anger, a critical, judgmental spirit, fear, avarice, greediness, controlling spirit, crude thoughts, uncertain sexual orientation. These may be distasteful traits that we criticize harshly in others, as we project them on to other people, forgetting that they are part of our makeup too. But they could be "golden shadows": undeveloped potential, underappreciated talents or dormant gifts, that we haven't been able to give space to in our life, because of time-restraints or fear of failure. We may admire these things in others but deny them in ourselves, feeling we couldn't possibly have the capacity to achieve such things. We have got stuck on the idea that such accomplishments are impossible for ourselves.

In the gospels, Jesus often says to his disciples at their safe, traditional hometown of Capernaum "Let us go over to the Other Side." The "Other Side" of the Sea of Galilee was regarded as enemy territory, the unclean Hellenistic land of the Decapolis, where lurk Gadarene demoniacs and swine—the eastern cliffs facing the lake finding themselves in shadow at the start of each day. Jesus' journey to the "dark side"—the other side of the lake—where he remained for significant periods and explored the territory, teaches us that we may embrace our shadow, and make friends with it: "You have heard that it was said, 'You shall love your neighbor and hate your enemy.' But I say to you, Love your enemies and pray for those who persecute you, so that you may be children of your Father in heaven; for he makes his sun rise on the evil and on the good, and sends rain on the righteous and on the unrighteous" (Matt 5: 43–45).

The "enemy" may be within: that shadow aspect of our personality that we detest in ourselves. Jesus commends a proper self-respect when he commands: "You shall love your neighbor as yourself" (Mark 12:31). We need to love ourselves and accept ourselves with the same kind of unconditional love that we see in Jesus as he embraces the prostitute, the tax collector, the thief on the Cross. We need to show to ourselves the kindness, gentleness, generosity and compassion that we would like to extend to others. Strategies are available that can help us embrace our shadow side.[11]

Removing the Veil

In the awesome theophany of Sinai, the Divine appears as fire: "Now the appearance of the glory of the Lord was like a devouring fire on the top of the mountain in the sight of the people of Israel" (Exod 24:17). The narrative

---

[11]. See, for example, Monbourquette, *Your Shadow*.

goes on to give us a memorable image about encountering God: the protective or concealing veil: "whenever Moses went in before the Lord to speak with him, he would take the veil off." Moses wore a veil to cover-over the shining on his face, which scared others, but it also represents potential barriers to God and to others. (Those who wear the Islamic Hijab today in western cultures are sometimes criticized about their concealment of the human face). Whatever the reason for putting it on, the point is that Moses took the veil off when he encountered God. There was no self-protective barrier, nothing to hide behind: rather a naked exposure to the Divine: "Thus the Lord used to speak to Moses face to face, as one speaks to a friend" (Exod 33:11; also Num 12:8;14:14).

This speaks to us of the strategies we resort to, the spiritual armor we grasp onto, in an attempt to keep God at a safe distance, where he may not singe or scorch us. We resort to endless strategies of avoidance and resistance, only limited by our ingenuity—we excuse ourselves from prayer, hide behind prayerbooks and holy beads, run out of time for silence, find that we are too busy, or too sleepy, for any encounter with the Divine. We tell ourselves that exposure to sun can be risky, and we should take care not to be make ourselves vulnerable to its penetrating rays. As Psalm 19:6 puts it: "Nothing is concealed from its [sun's] burning heat." The Book of Common Prayer says: "Almighty God, to whom all hearts be open, and from whom no secrets are hidden" (Collect for Purity, Holy Communion). In the encounter with God, we need to put *aside* sun-screens, sun-shades, and veils: "Now the Lord is the Spirit, and where the Spirit of the Lord is, there is freedom. And we all, with unveiled face, beholding the glory of the Lord, are being changed into his likeness from one degree of glory to another; for this comes from the Lord who is the Spirit" (2 Cor 3:17–18, RSV). And when we emerge from the time of prayer, we should be prepared to glow!

Mechthild of Magdeburg daringly develops the theme of spiritual nakedness before God. She relates a mystical dialogue with God who asks her:

> "Stay, Lady Soul."
> "What do you bid me, Lord?"
> "Take off your clothes."
> "Lord, what will happen to me then?"
> "Lady Soul, you are so utterly formed to my nature
> That not the slightest thing can be between you and me . . .
> You must cast off from you
> Both fear and shame . . ."
> "Lord, now I am a naked soul
> And you in yourself are a well-adorned God.
> Our shared lot is eternal life,

> *Without death."*
> Then a blessed stillness
> That both desire comes over them . . . [12]

Dare you approach the sun of the divine presence with utter openness? Dare you expose yourself to the divine fire? Will you take the risk of baring your soul to the penetrating rays of the Divine? Like sunflowers which orientate their faces to the sun as it passes across the sky, or like solar panels with receptor cells sensitive enough to catch every streak of sunlight and relay the energy received into a waiting grid, how ready are you to make yourself vulnerable, susceptible to God? As bodies crave the vital Vitamin D available from sunlight, how greatly do you long for the rejuvenating sun of God's grace?

Dare you re-enter the world of daily living with unveiled face, and be radiant, "aglow with the Spirit" (Rom 12:11)? Or will you cover up, for whatever reason?

God's Mission—and our Vocation

The Celtic tradition brings together the elemental metaphors speaking of the Divine and human:

> *God*
> I am the rays of the sun,
> I am the beam of the moon and the stars . . .
> *The soul*
> I am a flame of fire, blazing with passionate love;
> I am a spark of light, illuminating the deepest truth . . . [13]

Christians are called to be a sun in a dark world. Paul exhorts the Romans: "be aglow with the Spirit" (12:11). To Timothy he writes: "rekindle the gift of God that is within you" (2 Tim 1:6). In baptism the candidate is commissioned by the congregation in the words:

> You have received the light of Christ.
> Walk in this light all the days of your life.
> Shine as a light in the world, to the glory of God the Father!

This joyful commission echoes the words of Jesus in the gospel:

---

12. Tobin, *Mechthild*, 62.
13. From 'the Black Book of Carmarthen', de Weyer, *Celtic Fire*, 92.

The righteous will shine like the sun in the kingdom of their Father. (Matt 13:43)
In the same way, let your light shine before others, so that they may see your good works and give glory to your Father in heaven. (Matt 5:16)

The sun reminds us that we are called to be a consistent witness in a darkened world. We are called to reflect the light of God, to radiate to others his love. In John's gospel Jesus says: "I am the light of the world" (8:12). But in the Sermon on the Mount he says something different and startling. He says: "*You* are the light of the world!" (Matt 5:14)

QUESTIONS FOR REFLECTION

1. To what extent are you prepared to expose your soul to God? Do you take precautions—akin to sunscreen or sunshades—to prevent radical exposure and spiritual nakedness before God? In modern computing, users are urged to have in place a firewall, to prevent uninvited access to data or systems. Name the barriers or defensive strategies that you resort to, to keep the Divine at a safe distance. How might you remove or lower these barriers?

2. Have you experienced being dazzled by God's glory in any way? What is your experience of being filled with wonder or awe before the Divine?

3. Can you identify patches of darkness or shadow in your soul? Is anything obscuring or distorting the sunlight? What can you do about them?

4. Dare you allow yourself to be singed by the divine sun and fire, scorched by it? What ignites your spiritual life? Are you a sunny soul?

5. In what ways are you a beacon of hope and a shining light to those around you? How might you bring the light of Christ into someone's life today? Solar panels give us a contemporary image: every cell attuned and orientated to receive the maximum amount of sunlight, which is then passed on and energizes households. Ponder your baptismal vocation to "shine as a light in the world, to the glory of God the Father." Recall Christ's command: "let your light shine before others, so that they may see your good works and give glory to your Father in heaven" (Matt 5:16).

6. The sun is constant and consistent. Clouds come and go and obscure it. How consistent or variable is your Christian witness? Name anything clouding it!

PRAYER EXERCISE

Either

Practice a prayer that is first extravert in character, then introvert.

Begin by opening your arms wide. Let this bespeak utter exposure to God: open yourself to the penetrating rays of God's healing and invigorating grace. Stay in this mode as long as you can. Bask, as it were, in the sunlight of God's energizing love. What does it feel like? What is God saying to you?

Then, close your arms around your chest: let this speak to you of enclosure, being held by God. Feel enfolded and hemmed in by God's unconditional love. Permit yourself to be overwhelmed by God. Rest in this experience. Don't wriggle!

After the prayer time, make a review of the experience. What did utter exposure to the sunlight of God feel like for you? Describe possible feelings of vulnerability. What did the second way of praying feel like? What do you conclude—about yourself, and about God? End with the words of the psalm:

> Where can I go from your Spirit? Or where can I flee from your presence?
> If I ascend to heaven, you are there; if I make my bed in Sheol, you are there.
> If I take the wings of the morning and settle at the farthest limits of the sea,
>   even there your hand shall lead me, and your right hand shall hold me fast.
> If I say, "Surely the darkness shall cover me, and the light around me become night,"
>   even the darkness is not dark to you; the night is as bright as the day,
>   for darkness is as light to you. (Ps 139:7–12)

Or

Pray with some of the great hymns that celebrate Christ the Sun.

> I heard the voice of Jesus say, "I am this dark world's light;
> look unto me, thy morn shall rise, and all thy day be bright."
> I looked to Jesus, and I found in him my Star, my Sun;
> and in that light of life I'll walk till travelling days are done.
> (Horatio Bonar 1846)
>
> Sun of my soul, thou Savior dear, it is not night if thou be near;
> O may no earthborn cloud arise to hide thee from thy servant's eyes. (John Keble 1820)
>
> Jesus, Sun of righteousness, brightest beam of love divine,
> with the early morning rays do thou on our darkness shine,
> and dispel with purest light all our night . . .
> Like the sun's reviving ray, may thy love with tender glow
> all our coldness melt away, warm and cheer us forth to go,
> gladly serve thee and obey all the day.
> (Christian Knorr von Rosenroth, 1684; trans. Jane Borthwick, 1855)
>
> Summer suns are glowing over land and sea,
> happy light is flowing, bountiful and free.
> Everything rejoices in the mellow rays,
> all earth's thousand voices swell the psalm of praise.
>
> God's free mercy streameth over all the world,
> and his banner gleameth, everywhere unfurled.
> Broad and deep and glorious as the heaven above,
> shines in might victorious his eternal love.
>
> Lord, upon our blindness thy pure radiance pour;
> for thy loving-kindness make us love thee more.
> And when clouds are drifting dark across our sky,
> then, the veil uplifting, Father, be thou nigh.
>
> We will never doubt thee, though thou veil thy light:
> life is dark without thee; death with thee is bright.
> Light of Light! shine o'er us on our pilgrim way,
> go thou still before us to the endless day.
> (William Walsham How, 1871)

7

Entering the Cloud
Mystery and Presence

Sing to God, sing praises to his name;
  lift up a song to him who rides upon the clouds —
his name is the L<small>ORD</small> —
  be exultant before him . . .
You make the clouds your chariot
(Ps 68:4, 104:3)

Billowing with paradox, clouds are greeted with ambivalence and mixed feelings. We talk of someone who has "their head in the clouds" seeming out of touch with reality. Or we can be "on cloud nine" (if we can use a metaphor from the world of drugs). People say in exhilaration: "it feels like you're floating on air." William Wordsworth "wandered lonely as a cloud." But we can also confess "a heavy cloud hangs over us." Darkening clouds suggest the onset of conflict or trauma. We are invited to "blue sky thinking"—exploring new ideas in an unfettered way—and clouds might represent something that is obscuring or cluttering the creativity and luminosity of the spirit. Some might moan at the sight of clouds, asking "is it going to rain today?" But in places of drought the sight of a cloud is greatly welcomed as a sign of hope: "God commanded the skies above, and opened the doors of heaven"

(Ps 78:23). In some climes cloud may be resented, as blocking out precious sunlight, but even here we should be grateful, for clouds blanket a shivering earth in winter and replenish our reservoirs! Dando observes:

> The sky is an integral part of every landscape, and each cloud formation contains a message, and advanced notice of a potential change that impacts all who live in a particular place . . . Clouds were a component of the overall covenant of God with the ancient Israelites. Clouds reflected the power, wisdom, presence, and promises of God.[1]

Sometimes we need to learn see things differently. For example, the "dark night of the soul" turns out, in the writings of John of the Cross, to be a potentially positive time in which to experience God ever more deeply. The desert experience of wilderness-time can be, for Jesus in the gospels and for the desert fathers and mothers, hugely creative and formative, a time for glimpsing new visions and clarifying priorities. In this chapter we discover the cloud paradoxically to be both offering promise and warning . . .

Scudding across the thirsty terrain of the Holy Land, clouds appear as harbingers of promise. They represent hopefulness for renewal. Jesus speaks of a "cloud rising in the west" (Luke 12:54) referring to the rain-bearing cumulus coming in from across the Mediterranean. Jesus recalled the ministry of Elijah in a time of desperate need "when the heaven was shut up for three years and six months, and there was a severe famine over all the land" (Luke 4:25). Elijah had awaited the advent of moisture-laden clouds from the sea, as they gradual build in density and size. This required a sevenfold weather check:

> Elijah said to Ahab, "Go up, eat and drink; for there is a sound of rushing rain." So Ahab went up to eat and to drink. Elijah went up to the top of Carmel; there he bowed himself down upon the earth and put his face between his knees. He said to his servant, "Go up now, look towards the sea." He went up and looked, and said, "There is nothing." Then he said, "Go again seven times." At the seventh time he said, "Look, a little cloud no bigger than a person's hand is rising out of the sea." Then he said, "Go and say to Ahab, "Harness your chariot and go down before the rain stops you." In a little while the heavens grew black with clouds and wind; there was heavy rain. Ahab rode off and went to Jezreel. But the hand of the Lord was on Elijah; he girded up his loins and ran in front of Ahab to the entrance of Jezreel. (1 Kgs 18:41–45)

1. Dando, "Clouds," 61, 65.

Ahab had worshipped Baal, the lord of the rainclouds, but it was the King of heaven that answered the prayer.

CLOUDS OF THE OLD COVENANT

Clouds on the Journey

The Hebrews first sensed the presence of God in the cloud in their experience of the Exodus journey to liberation:

> The Lord went in front of them in a pillar of cloud by day, to lead them along the way, and in a pillar of fire by night, to give them light, so that they might travel by day and by night. Neither the pillar of cloud by day nor the pillar of fire by night left its place in front of the people. (Exod 13:21,22)

It not only represented a sign of God's abiding presence among his people, it also shifted its location as if to guide and protect the people:

> The angel of God who was going before the Israelite army moved and went behind them; and the pillar of cloud moved from in front of them and took its place behind them. It came between the army of Egypt and the army of Israel. And so the cloud was there with the darkness, and it lit up the night; one did not come near the other all night... At the morning watch the Lord in the pillar of fire and cloud looked down upon the Egyptian army, and threw the Egyptian army into panic. (Exod 14:19,20,24)

The mysterious event of the theophany atop Mount Sinai is vividly described:

> Then the Lord said to Moses, "I am going to come to you in a dense cloud, in order that the people may hear when I speak with you and so trust you ever after."... Then Moses went up on the mountain, and the cloud covered the mountain. The glory of the Lord settled on Mount Sinai, and the cloud covered it for six days; on the seventh day he called to Moses out of the cloud ... Moses entered the cloud, and went up on the mountain. Moses was on the mountain for forty days and forty nights. (Exod 19:9,16; 24:15,16,18)

The cloud was closely associated with the Tabernacle, the Tent of Meeting, which housed the travelling Ark of the Covenant:

> When Moses entered the tent, the pillar of cloud would descend and stand at the entrance of the tent, and the Lord would speak

with Moses. When all the people saw the pillar of cloud standing at the entrance of the tent, all the people would rise and bow down, all of them, at the entrance of their tents . . . The Lord descended in the cloud and stood with him there, and proclaimed the name, "The Lord . . . ." Then the cloud covered the tent of meeting, and the glory of the Lord filled the tabernacle . . . Whenever the cloud was taken up from the tabernacle, the Israelites would set out on each stage of their journey; but if the cloud was not taken up, then they did not set out until the day that it was taken up. (Exod 33:9,10; 34:5; 40:34,36,37)

The cloud appears as a re-assuring but paradoxical symbol of the Divine. Its presence was both comforting and scary. The cloud is almost palpable but enigmatic. It certainly does not stay static for long—it is on the move, dynamic. It represents the pilgrim God, a God going places with his people. The cloud at once signifies the nearness of God but also the fact that he is uncontainable, elusive, incapable of being boxed in or tied down. At the same time it denotes the guidance of God and his unpredictability.

Clouds in the City

When in about 970 BC Solomon builds his Temple, the presence of the Divine is not only represented in the static Ark, but in the mysterious cloud that comes and goes, and varies in density and intensity. Chronicles relates:

> The house, the house of the Lord, was filled with a cloud, so that the priests could not stand to minister because of the cloud; for the glory of the Lord filled the house of God. Then Solomon said, "The Lord has said that he would reside in thick darkness. I have built you an exalted house, a place for you to reside in for ever" (2 Chr 5:13–6:2)

Later, rabbis would explicate this in terms of the *shekinah* glory of God—the Semitic roots of the word mean "to settle, inhabit, or dwell." There was some sense that God manifested or expressed his glorious presence in the cloud. But the paradox remains, expressed by Solomon himself: "But will God indeed dwell on the earth? Even heaven and the highest heaven cannot contain you, much less this house that I have built!" (1 Kgs 8:27).

The psalms, hymnbook of the Temple, go on to recall Sinai and to welcome God's misty presence in the sanctuary of Zion, celebrating both the darkness and brightness of God's clouds:

> Then the earth reeled and rocked;
>> the foundations also of the mountains trembled
>> and quaked, because he was angry.
> Smoke went up from his nostrils,
>> and devouring fire from his mouth;
>> glowing coals flamed forth from him.
> He bowed the heavens, and came down;
>> thick darkness was under his feet.
> He rode on a cherub, and flew;
>> he came swiftly upon the wings of the wind.
> He made darkness his covering around him,
>> his canopy thick clouds dark with water.
> Out of the brightness before him
>> there broke through his clouds
>> hailstones and coals of fire. (Ps 18:7–12)

These perceptions are echoed by Job. Humbled before the mystery of God, he poses unanswerable questions:

> Can anyone understand the spreading of the clouds,
>> the thunderings of his pavilion? ...
>>> Do you know the balancings of the clouds,
>> the wondrous works of the one whose knowledge is perfect ...
> ?
>>> Who has the wisdom to number the clouds?
>> Or who can tilt the waterskins of the heavens ... ?
> (Job 36:29; 37:16; 38:37)

Clouds, then, summon us to wonderment and silence before the Divine. They represent something we cannot work out, something we can't control, something sovereign, enigmatic, elusive ...

Metaphors of the Spirit

Not only do the scriptures depict God's being through the imagery of clouds—they also glimpse in the skies clues and pointers about the human condition. On the one hand, clouds might be a symbol of promise and hope, indeed represent the potential goodness of a leader:

> In the light of a king's face there is life,
>> and his favor is like the clouds that bring the spring rain.
> (Prov 16:15)

But what of waterless clouds that don't deliver? They become a sign of emptiness, broken promises, disappointment:

> Like clouds and wind without rain
>   is one who boasts of a gift never given. (Prov 25:14)

They can speak of lives that look satisfactory on outside but in fact are lacking, clouds that have no moisture to deliver to a thirsty planet. In the Holy Land, waterless clouds, the high cirrus clouds drawing in desert air from the south and east borne by sirocco wind, furnish both Jude and Peter with imagery to describe fickle human beings:

> They are waterless clouds carried along by the winds; autumn trees without fruit, twice dead, uprooted; wild waves of the sea, casting up the foam of their own shame; wandering stars, for whom the deepest darkness has been reserved for ever. (Jude 1:12,13)
> These [people] are waterless springs and mists driven by a storm; for them the deepest darkness has been reserved. For they speak bombastic nonsense, and with licentious desires of the flesh they entice people who have just escaped from those who live in error. (2 Pet 2:16–18)

These comparisons vividly bring out the dry, empty and purposeless existence of such people. Morning clouds that quickly evaporate become symbolic of transitory things, and of people who are unreliable, with passing, fleeting, devotion:

> What shall I do with you, O Ephraim?
>   What shall I do with you, O Judah?
> Your love is like a morning cloud,
>   like the dew that goes away early. (Hos 6:4, cf. 13:3)
> As the cloud fades and vanishes,
>   so those who go down to Sheol do not come up. (Job 7:9)

Job himself interprets his passing possessions and disappearing resources through such imagery:

> Terrors are turned upon me;
>   my honor is pursued as by the wind,
>   and my prosperity has passed away like a cloud. (Job 30:15)

CLOUDS OF THE NEW COVENANT

At the baptism of Jesus the heavens were "torn apart." But it is in the awesome events of the transfiguration and ascension that the biblical writers direct our attention to the clouds. "Then a cloud overshadowed them, and from the cloud there came a voice, 'This is my Son, the Beloved; listen to him!'" (Mark 9:7). What is the significance of the cloud at this pivotal moment in the life of Jesus?

Cloud over Tabor

Let's notice at what point in the story the cloud appears. Peter shouts out: "Let us make three dwellings" (Mark 9:4). What were Peter's motives in this plan to build three tents or tabernacles? Often in time of fear or stress one's true nature comes out. It seems that Peter wanted to regain mastery over circumstances which were unfolding without his scheming. He had lost the initiative. He needed to take charge. Perhaps, also, he needed to box in the mystery of the appearances that was too great for his mind to grasp—he needed to get a handle on things, as we say. He needed to be able to take a grip on the situation, so he suggested that he would construct shelters to hold things in. Peter's heart quakes with fear of the unknown. But above all, he simply needed to be in control!

What is God's response to his request? It comes in the form of a dense mist or fog that envelops and enshrouds them: "Then a cloud overshadowed them . . . they entered the cloud" (Luke 9:34). The divine mystery was too big for any kind of tent! This can represent our own attempts in prayer to encase the untamable Divine with words, concepts and images. But precisely at the very point when Peter suggests the building of booths "while he was saying this a cloud came and overshadowed them; and they were afraid as they entered the cloud" (Luke 9: 34). The response to human tent-building is a divine smothering or drenching in mysterious wet mist where visibility is reduced to nil. The cloud now dampens the senses and exuberant conceptualizing and silences the over-active mind. The cloud eclipses the sun: there has been, as it were, a change in the weather, from bright sunlight to darkening cloud, gloom and impenetrable haze. A swirling fog blankets the disciples. This expresses well the sense of disorientation and confusion that one experiences in a baffling liminal place. In the cloud one feels out of control, not knowing which way to turn. It turns out to be a poignant symbol of that transition in prayer from active, discursive thinking to simpler loving.

On the mount of transfiguration Jesus encounters God in both the light and the cloud. The light represents the *kataphatic* tradition of prayer—where affirmations and declarations about God are made confidently: Jesus is light; "his face shone like the sun"; "his garments became white as light." But the cloud represents the *apophatic* tradition, where words give way to silence, where concepts about God dissolve into speechless wonder, where the unifying dense wet fog of the cloud shrouds the disciples and silences all attempts at talking. The reference to the overshadowing cloud evokes the theophany on Mount Sinai, where God appears in "a thick cloud on the mountain" (Exod 19:16). In his *Life of Moses*, Gregory of Nyssa (330–395) tracing the Christian path on the analogy of the Exodus, tells us that there are times when we need to allow ourselves to be overshadowed by the cloud of God's utter mystery:

> For leaving behind everything that is observed, not only what sense comprehends but also the intelligence thinks it sees, it keeps on penetrating deeper until by the intelligence's yearning for understanding it gains access to the invisible and the incomprehensible, and there it sees God. This is the true knowledge of what is sought; this is the seeing that consists in not seeing, because that which is sought transcends all knowledge, being separated on all sides by incomprehensibility as by a kind of darkness.[2]

The classic fourteenth century text *The Cloud of Unknowing* invites us to experience the cloud of prayer. The author pursues a similar apophatic approach, the *via negativa*, cautious about the use of vivid images in relation to speaking of God, commending wordless silence rather than exuberant speech:

> When you first begin, you find only darkness and as it were a cloud of unknowing. You don't know what this means except that in your will you feel a simple steadfast intention reaching out towards God . . . Reconcile yourself to wait in this darkness as long as is necessary, but still go on longing after him whom you love. For if you are to feel him or to see him in this life, it must always be in this cloud, in this darkness.[3]

The *Cloud* pays attention to the affective aspects of prayer, rather than the cognitive aspects. It says of God: "He may well be loved, but not thought.

---

2. Malherbe & Ferguson, *Gregory of Nyssa*, 95.

3. Wolters (tr.), *Cloud*, 53,54. For a more recent translation see Walsh (tr.), *Cloud*. See also Lonsdale, "The Cloud of Unknowing."

By love he can be caught and held, but by thinking never."[4] The *Cloud* aids the identification of "signs" of spiritual progress or impediments to growth. It teaches that there are various signs, clues or evidences that suggest that the reader may be ready to make a transition in their praying from discursive, active thinking with words and images, as in meditations on the passion, towards the wordless silence and solitude of contemplation. One key indicator is that of desire or yearning: "our Lord in his great mercy has called you ... leading you on to himself by your heart-felt desire."[5] The author wants to urge the Christian to step forward into a transfiguration-kind of prayer characterized by watching, waiting, longing . . . "So when you feel by the grace of God that he is calling you to this work, and you intend to respond, lift up your heart to God with humble love . . . It all depends on your desire."[6]

There are other clues or indicators that suggest that someone may be ready to enter more mystical, receptive prayer. One finds oneself unfulfilled by Peter's tents, representing prayer hemmed in by words. Even liturgical prayer or the daily office or traditional devotions come to feel constricting and stifling. The Cloud beckons.

Light and Cloud

We glimpse other vital lessons in the transition from dazzling sunlight to darkening cloud. Jesus "went up the mountain to pray." Luke offers us this episode to teach us something about prayer. The first part of the transfiguration story, the radiant light surrounding Christ, speaks to us of the times it is easy to pray: the light is shining. In bright times we can easily glimpse the glory and presence of God—in the beauty of creation, the playfulness of children, the smile of a friend or stranger. But we also face times when life is foggy, cloudy—when we can't see the way forward, when we find ourselves in a haze of confusion or indecision, or when the dark clouds of suffering, anxiety or illness hang overhead. We find ourselves in a cloud of struggle. The story of the transfiguration assures us: God is in the light, but he is in the dark cloud too—indeed he speaks to us from the cloud, calling to us "Listen!" We are to discern his presence in the shadows, and try to hear what he is saying to us in them. Then we find ourselves, as Matthew puts it, in "a bright cloud" (Matt 17:5).

---

4. Wolters, *Cloud*, 60.

5. Wolters, *Cloud*, 51. On the role of desire in spirituality see Sheldrake, *Befriending Our Desires*.

6. Wolters, *Cloud*, 61.

Jesus in the Cloud

The Acts of the Apostles gives us an account of the ascension: "as they were watching, he was lifted up, and a cloud took him out of their sight" (1:9). The cloud that receives Jesus in his exultation bespeaks the impenetrable and unfathomable mystery of God. But it is also tantalizing and intriguing. It becomes a sign of hope:

> Then they will see "the Son of Man coming in clouds" with great power and glory. (Mark 13:26)
> Look! He is coming with the clouds; every eye will see him. (Rev 1:7)

For his part, Paul is confident:

> For the Lord himself, with a cry of command, with the archangel's call and with the sound of God's trumpet, will descend from heaven, and the dead in Christ will rise first. Then we who are alive, who are left, will be caught up in the clouds together with them to meet the Lord in the air; and so we will be with the Lord for ever.

And he concludes:

> Therefore encourage one another with these words. (1 Thess 4:16–18)

The cloud endures as a great symbol of expectation, indeed a sign of our destiny. It is distant yet beckoning, mysterious yet inviting. And we don't forget that, in the experience of prayer, we are invited with Peter to enter the cloud, here below.

QUESTIONS FOR REFLECTION

1. What does the symbol of the cloud speak to you about the nature of God and about human nature? Do you have any changing perceptions about the cloud—maybe seeing it differently now?
2. In what ways can clouds be a mirror of your soul?
3. As you reflect on your own personal faith history, where have you discovered moments of transfiguration or transformation?

4. What is your experience of "praying in the cloud" or the "apophatic mode" of prayer as tradition calls it? What do you make of "the cloud of unknowing"?

5. Are you tempted, like Peter, to build "tents" in your prayer-times? What do they symbolize for you?

6. Joel writes: "the day of the Lord is coming, it is near— a day of darkness and gloom, a day of clouds and thick darkness!" (Joel 2:1–2). In our time we have witnessed the terrifying mushroom cloud of Hiroshima, and the clouds of radioactive pollution arising from Chernobyl. What dark clouds overshadow us today? How do you find yourself responding to them, and what kind of prayer do you offer?

PRAYER EXERCISE

Either

Revisit the account of the transfiguration (Matt 17 or Luke 9:28–36) to meditate on it in an Ignatian way. Ignatius says: Use your eyes to *look* at the scene, visualize it, imagine it in your mind's eye, place yourself into the picture and become one of the characters. Reach out in your imagination and *touch* with your fingertips the characters, the soil, the water, the physical aspects. Even *smell* the scents of the scene and *taste* the air, the atmosphere. But above all, Ignatius says, open your ears and *listen* to what the characters are saying to each other, what they are saying to you and what God is saying to you through all this.

Picture yourself joining Peter and James and John in their ascent of the holy mountain. As you go through the story step by step, notice how you find yourself reacting—to the dazzling light or becoming enshrouded by the dense cloud. What does that feel like? Conclude with Ignatius' own prayer: "Take, O Lord, and receive my entire liberty, my memory, my understanding and my whole will. All that I am and all that I possess You have given me: I surrender it all to You to be disposed of according to Your will. Give me only Your love and Your grace; with these I will be rich enough."

Or

This prayer-time is in two phases.

First, sit for a while in utter darkness. Let the darkness and silence speak to you of people's longing for God—their deep need for Christ's revelation. Also, as you quieten your heart and silence your lips, pray in the *apophaptic* mode—with wordless wonder and no attempt at describing the Divine.

Secondly, when you are ready, light a candle before you. See how the light dispels the darkness. Look at the flame and find yourself praying that you will be a radiant light in the world, revealing the wonder and mystery of Jesus to others. Pray now in the *kataphatic* mode—affirming your love for Christ, and attempting to find words to express your wonder. Pray that your light may intensify and burn ever brighter as you yourself discover more of him. Pray that epiphany may take place during your life today.

8

Thundering in the Soul
Protest and Silence

The voice of the Lord is over the waters;
    the God of glory thunders,
    the LORD, over mighty waters.
The voice of the LORD is powerful;
    the voice of the LORD is full of majesty.
        The voice of the LORD breaks the cedars;
    the LORD breaks the cedars of Lebanon . . .
The voice of the LORD flashes forth flames of fire.
The voice of the LORD shakes the wilderness;
    the LORD shakes the wilderness of Kadesh.
        The voice of the LORD causes the oaks to whirl,
    and strips the forest bare;
    and in his Temple all say, "Glory!"
        The LORD sits enthroned over the flood;
    the LORD sits enthroned as king for ever.
May the LORD give strength to his people!
    May the LORD bless his people with peace!

(Ps 29)

Crashing, rolling, booming, roaring, growling: the sounds of thunder reverberate across the pages of scripture and vibrate within the soul. When the sound of thunder rumbles overhead we might quake or shiver. Roof-slates loosen, crockery rattles, dogs bark. We find ourselves quite naturally using phrases like "he stole my thunder," "a face like thunder" and "the thunder of protest." What will we discover in the Bible?

A Voice from Heaven

The Bible begins with the creative *dabar*, the voice from heaven: "Then God said, 'Let there be light'; and there was light" (Gen 1:3). When the divine voice thunders across the heaving chaos, things start to happen—in the poem of Genesis 1, God's every word brings forth fresh creations from nothing.

At the end of the Bible, the divine rumbling is heard again. The seer of the Book of Revelation relates: "I heard a voice from heaven like the sound of many waters and like the sound of loud thunder; the voice I heard was like the sound of harpists playing on their harps" (14:2). Indeed, "seven thunders" pound the earth:

> And I saw another mighty angel coming down from heaven, wrapped in a cloud, with a rainbow over his head; his face was like the sun, and his legs like pillars of fire. He held a little scroll open in his hand. Setting his right foot on the sea and his left foot on the land, he gave a great shout, like a lion roaring. And when he shouted, the seven thunders sounded. And when the seven thunders had sounded, I was about to write, but I heard a voice from heaven saying, "Seal up what the seven thunders have said, and do not write it down." Then the angel whom I saw standing on the sea and the land raised his right hand to heaven and swore by him who lives for ever and ever, who created heaven and what is in it, the earth and what is in it, and the sea and what is in it: "There will be no more delay, but in the days when the seventh angel is to blow his trumpet, the mystery of God will be fulfilled, as he announced to his servants the prophets." (Rev 10:1–7)

But while this sounds forbidding, joyful praises sound forth too: "Then I heard what seemed to be the voice of a great multitude, like the sound of many waters and like the sound of mighty thunder-peals, crying out, 'Hallelujah! For the Lord our God the Almighty reigns'" (19:6).

The primal, primordial sound of thunder evokes the awesome accounts of the theophany on Mount Sinai:

> On the morning of the third day there was thunder and lightning... and a blast of a trumpet so loud that all the people who were in the camp trembled. Moses brought the people out of the camp to meet God. They took their stand at the foot of the mountain. Now Mount Sinai was wrapped in smoke, because the Lord had descended upon it in fire; the smoke went up like the smoke of a kiln, while the whole mountain shook violently. As the blast of the trumpet grew louder and louder, Moses would speak and God would answer him in thunder. When the Lord descended upon Mount Sinai, to the top of the mountain, the Lord summoned Moses to the top of the mountain, and Moses went up. (Exod 19:16–24)

This becomes the archetypal symbol of thunder. It denotes the revelation of God, God making his will known to men and women, and God seeking a covenant relationship with them. The Psalms tremble with a sense of wonderment:

> The Lord also thundered in the heavens,
>   and the Most High uttered his voice. (Ps 18:13)
> The crash of your thunder was in the whirlwind;
>   your lightnings lit up the world;
>   the earth trembled and shook. (Ps 77:18)

Wisdom literature celebrates the divine voice:

> His voice roars;
>   he thunders with his majestic voice
>   and he does not restrain the lightnings when his voice is heard.
> God thunders wondrously with his voice;
>   he does great things that we cannot comprehend. (Job 37:4–5)

The prophets have the grave responsibility to mediate the divine voice to humans. Jeremiah is given this charge to convey:

> The Lord will roar from on high,
>   and from his holy habitation utter his voice;
> he will roar mightily against his fold,
>   and shout, like those who tread grapes,
>   against all the inhabitants of the earth. (Jer 25:30)

Longing for the Divine Voice

After the decline of prophecy, ending with Malachi, there was a great longing to catch the echo of God's voice, the voice from heaven. A nostalgia for Sinai mingled with a yearning for Zion. The references to the "voice from heaven" in the gospels need to be set against this background of longing, articulated by the Jewish sages known as the Tannaim, who were active 10-200 AD and recorded their thoughts in the Mishnah, the rabbinical collection of oral traditions. There we read of the concept of the *bath'-kol*, "the daughter of the voice." This phrase derives from Daniel: " While the words were still in the king's mouth, a voice came from heaven: 'O King Nebuchadnezzar, to you it is declared: The kingdom has departed from you!'" (Dan 4:31). In the period of the Tannaim the term *bath'-kol*, was in very frequent use, understood to signify not the direct voice of God, held to be beyond the senses, but the echo of the voice. The rabbis held that *bath'-kol* had been an occasional means of divine communication throughout the whole history of Israel and that since the cessation of the prophetic gift it was the sole means of divine revelation:

> The Lord thundered from heaven;
> the Most High uttered his voice. (2 Sam 22:14)

Thunder Returns

Such a sense of expectancy lies in the background of the gospel accounts of the baptism and transfiguration of Jesus. Mark relates:

> In those days Jesus came from Nazareth of Galilee and was baptized by John in the Jordan. And just as he was coming up out of the water, he saw the heavens torn apart and the Spirit descending like a dove on him. And a voice came from heaven, "You are my Son, the Beloved; with you I am well pleased." (Mark 1:9-11)

Jesus receives a profound affirmation of his identity as he emerges, dripping, from the waters of the Jordan. This will be repeated on the mountain of revelation, evoking Sinai itself, as we began to see in the last chapter. Peter's own account of the transfiguration highlights the experience of the booming Voice from heaven: "That voice was conveyed to him by the Majestic Glory, saying, 'This is my Son . . . ' We ourselves heard this voice come from heaven, while we were with him on the holy mountain" (2 Pet 1:17, 18).

Luke suggests that the prayer experience of the transfiguration is a place of awakening, heightened consciousness and alert awareness: "Now

Peter and his companions were weighed down with sleep; but when they were fully awake, they saw his glory" (Luke 9:32). The disciples are awakened, dazzled, awestruck. They will never be the same again. They are invited to a wakefulness in which their spiritual senses are put on high alert. They are invited to *look*: to gaze on the mystery, to open their eyes to the light. They are invited to open their ears and *listen*: the Father's voice, from heaven, calls out: "Listen to him!" They are *touched*: the moisture of the wet mist soaks their skin, and Jesus reaches out his hands to them to touch them and lift them up. There is an awakening of the spirit and the body: a coming fully-alive, aware and responsive to what God wanted to offer them in this prayer experience. Bonaventure writes that prayer requires the rediscovery of the spiritual senses: "when the inner senses are restored to see the highest beauty, to hear the highest harmony, to smell the highest fragrance, to taste the highest sweetness, to apprehend the highest delight, the soul is prepared for spiritual ecstasy through devotion, admiration and exultation."[1] This is the invitation of prayer.[2]

Finally, the divine thunder rumbles in John's account of the passion:

> "Now my soul is troubled. And what should I say—"Father, save me from this hour"? No, it is for this reason that I have come to this hour. Father, glorify your name." Then a voice came from heaven, "I have glorified it, and I will glorify it again." The crowd standing there heard it and said that it was thunder. Others said, "An angel has spoken to him." Jesus answered, "This voice has come for your sake, not for mine. Now is the judgment of this world; now the ruler of this world will be driven out. And I, when I am lifted up from the earth, will draw all people to myself." He said this to indicate the kind of death he was to die. (John 12:27–32)

John gives us a Gethsemane-like scene. Westcott comments: "The utterance was real and objective, that is, it was not a mere thunder-clap interpreted in this sense; yet, like all spiritual things, this voice required preparedness in the organ to which it was addressed."[3] Yes—God is thundering from heaven—"look at the Cross—things will never be the same again."

---

1. Cousins, *Bonaventure*, 89.

2. Ignatius of Loyola invites us to use our five senses to trigger our imaginations as we engage with the text: Mottola, *Spiritual Exercises*.

3. Westcott, *St John*, 182.

Sons of Thunder

The Galilee of Jesus' time suffered the double trouble of oppression and poverty. The Galileans were crippled by heavy taxes: dues were owed to the Roman occupier, and Temple taxes added to the burden. At the time of Jesus ordinary families were being forced to quit their ancestral landholdings, where they had lived for centuries, in order to meet these demands. But then they had to pay rent for what had been their own fields and homes: they became caught in a downwards economic spiral, becoming tenants in their own property: many of Jesus' parables speak of absentee landlords who impose severe dues on their tenants (see, for example, Luke 16:1–8; Matt 25:14–30). Tax and rent robbed the Galilean peasant farmer of two thirds of the family income. Many were living at barely subsistence level. No doubt Matthew preserves an original aspect of the Lord's Prayer when he puts it: "Forgive us our debts and we forgive those who are in debt to us" (Matt 6:12). It was a world of "haves" and "have nots."

It is against this background that we meet "the Sons of Thunder." At the time of Jesus, two expressions of resistance and protest against the status quo emerged. First, there were the terrorists. "Have you come out with swords and clubs to arrest me as though I was a bandit?" (Mark 14:48). The Greek word *lestes*, translated "bandit," denotes freedom fighter or even terrorist. Josephus tells us about revolutionary activists based in Galilee who sought to undermine Roman domination by acts of sabotage. Since the revolt of Judas the Galilean in 4 BC the region had become a hotbed of resistance to increasingly stifling imperial rule.[4] In the accounts of the passion, it is clear that Jesus is on trial before the Roman authorities precisely for being a subversive social revolutionary. He is arraigned beside the insurrectionist Barabbas. He is crucified between two "thieves"—this is the usual tame translation (Mark 15:27), but the Greek word "bandit" reminds us of this growing movement of rebellion and insurgency against Roman oppression.

The second group we encounter at this time were the Zealots. The hand-picked band of disciples included Judas Iscariot—his surname may relate to the Sicarri rebels, forerunners of the Zealots. Also we meet "Simon the Zealot" and the brothers "James son of Zebedee and John the brother of James, to whom he gave the name Boanerges, that is, Sons of Thunder" (Mark 3:17). This nickname may point to their involvement in this growing protest movement against Roman oppression.[5] As we noted, Ecclesiastes 3

---

4. See Bammel & Moule, *Jesus and the Politics of his Day*.
5. It could also refer to their grumpy father or feisty mother!

reminds us: there is "a time to keep silence, and a time to speak." Thunderous anger can be righteous anger in the face of dehumanizing conditions.

Maybe, therefore, a third of the Twelve were involved one way or another in the protest movement that was raising steam at the time of Jesus. Peter Walker writes: "The Palestine in which Jesus grew up was politically red-hot ... The tension between the Jews and Roman rulers was increasing. Jesus found himself in a context that was like a tinderbox waiting to go up in flames."[6]

THUNDER AND SILENCE

Between lightning and thunder there is a silence. It is an expectant silence, for in it we listen out in anticipation, wondering how long it will be, after the lightening flash, until we hear the crashing of thunder. Of course, this tells us if the storm is overhead, as light travels faster than sound. In the silence we find ourselves waiting, watching, listening intently. We need to hold in a creative tension and dialectic the crashes of thunder and the sounds of silence. The story of Elijah reminds us that God can be found in silence:

> He got up, and ate and drank; then he went in the strength of that food for forty days and forty nights to Horeb the mount of God. At that place he came to a cave, and spent the night there. Then the word of the Lord came to him, saying "Go out and stand on the mountain before the Lord, for the Lord is about to pass by." Now there was a great wind, so strong that it was splitting mountains and breaking rocks in pieces before the Lord, but the Lord was not in the wind; and after the wind an earthquake, but the Lord was not in the earthquake; and after the earthquake a fire, but the Lord was not in the fire; and after the fire a sound of sheer silence. When Elijah heard it, he wrapped his face in his mantle and went out and stood at the entrance of the cave. Then there came a voice to him that said, "What are you doing here, Elijah?" (1 Kings 19: 8–14).

Dorothy Soelle (1929–2003) in *The Inward Road and The Way Back* reflects on the experience of Elijah as related in 1 Kings 19 and his encounter with God, not in earthquake, wind or fire but in "the still small voice." She observes that Elijah did not linger there, for the still small voice uttered a political charge: "Then the Lord said to him, 'Go, return on your way to the wilderness of Damascus; when you arrive, you shall anoint Hazael as king over Aram ... '" (1 Kings 19:15). Soelle points out:

6. Walker, *Jesus and His World*, 30.

[after] the experience of God in the "still, small voice" what happens now? Elijah does not withdraw into an act of worship; he does not make a pilgrimage to some shrine. Nor does he continue to divide things into the categories of sacred and profane, a division so dear to all religions. Instead, what happens is of significance for the Judeo-Christian tradition: the renewal of his political mission . . . he returns to the world.[7]

Elijah's waiting in the silence after the crashing noise leads to speech and action. Prayer that listens to God with attentiveness can thus become transformative, changing and reshaping our priorities and plans. Jesus models for us this balance between the mystical and the prophetic, between silence and speaking, between action and contemplation. We saw in chapter three how Jesus speaks out courageously, decisively and prophetically when he finds himself in the midst of the storm of controversy. When needed he thunders with righteous anger. But his authority, message and courage spring from his times of expectant silence.

Jesus exemplifies this dialectic between thunder and silence. In Mark's gospel, the Twelve are chosen (ch.3) "to be with him and to be sent out." They are to spend quality time in the presence of Jesus and then venture forth in their apostolate. Both Mark and Luke emphasize the role of prayer and silence in the example Jesus sets before the disciples, following the forty days of prayer, struggle and preparation in the desert prior to the start of his public ministry. In Mark chapter 1, a hectic twenty-four hours of ministry is followed by prayer before dawn in an *eremos*—lonely place (1:35): the time of prayer is both the conclusion of an intense period of ministry and the prelude to the next stage. This rhythm of prayer and activity is repeated in the disciples' experience, as they go to a place of retreat enabling rest and reflection after first incursions into ministry and giving an account to Jesus (Mark 6:30,31). After this retreat, another time of ministry (6:35–45) is followed by Christ's retirement into the hills for prayer at night (6:46): the pattern of intense activity and solitude is repeated. As Dunn puts it, we should note "the degree to which Jesus provided a model to his disciples as a man of prayer . . . To be a disciple of Jesus was to pray as Jesus prayed."[8]

While Jesus cherishes and safeguards times of aloneness, he also brings his stillness into the midst of the noisy world: his desert heart still pulsates within him. But he must leave the lonely places—heartened, challenged, instructed, comforted and energized to face the demands of ministry and the call of the Cross. This rhythm between withdrawal and engagement,

7. Soelle, *Inward Road*, 136.
8. Dunn, *Jesus Remembered*, 561.

this ebb and flow of prayer and ministry, is the key to the ministry of Jesus: he moves with a listening heart amidst a clamoring, demanding world. The Jesus of John's gospel can only share and reveal what he himself has heard from his Father: "He testifies to what he has seen and heard, yet no one accepts his testimony" (3:32). Jesus is emphatic: "the one who sent me is true, and I declare to the world what I have heard from him" (8:26). He describes himself as "a man who has told you the truth that I heard from God" (8:40). He is clear: "the word that you hear is not mine, but is from the Father who sent me" (14:24). As he speaks, he is listening!

The greatest challenge is not only to set aside alternating times for prayer and stillness and times for service. It is to bring a contemplative heart into the bustling center of ministry. Jesus models not only the ebb and flow of prayer and action, but also the ability to maintain a listening heart in the very maelstrom of ministry. Certainly he lives within a rhythm of withdrawal and engagement, but it is in the heat of fierce debate that he is able to say: "Very truly, I tell you, the Son can do nothing on his own but only what he sees his Father doing . . . The Father loves the Son and shows him all that he himself is doing" (John 5:19,20). Thunder and silence are not far apart!

THUNDER IN THE SOUL

In his great poem "Prayer I" George Herbert describes prayer as

> *The soul in paraphrase, heart in pilgrimage,*
> *The Christian plummet sounding heaven and earth;*
> > *Engine against the Almighty, sinner's tower,*
> > *Reversed thunder . . .*

Prayer is the place where we can bring to God the trials, joys, questions and paradoxes of life. *Engine against the Almighty*—referring to a siege engine—represents the act of throwing at God our feelings. *Reversed thunder* suggests that we can mirror the weather in our soul. Thunderbolts descend from heaven to earth, but in prayer we can direct heavenwards and Godwards our most intense feelings. We come before God in our fragility and vulnerability, and in our brokenness, and we do not conceal from him our woes and heartaches, nor do we hold back from throwing at God our most perplexing questions and frustrations.[9] The ancients thought that thunder was produced by clouds banging together in collision. This speaks to us of those times when our soul is stormy, when peace is missing, when things clash or compete—times of spiritual upheaval and conflict. Herbert shares

---

9. Herbert, *English Works*; Hutchinson, *Works of George Herbert*.

in his poems an anguishing cry of the heart, expressed in searing honesty and revealing a deep authenticity of soul. The thunder of soul is unavoidable and must be expressed, as in nature, when lightning flashes, thunder follows.

The Anglican poet-priest of the seventeenth century, George Herbert wrote that his poems were "a picture of the many spiritual conflicts that have passed betwixt God and my soul."[10] His poems testify to an on-going struggle to accept personally within himself God's unconditional love. Herbert was born in 1593 to an aristocratic family. After studies at Cambridge University he became a lecturer in Rhetoric and for seven years held the prestigious post of Public Orator to the university. He seemed destined for high office, and set his hopes on a privileged career in the royal court, but God had other plans for him. Secular ambitions wrestled with a persistent and nagging sense of vocation to the priesthood, and Herbert finally gave in and was ordained deacon in 1626. But things were not to be straightforward for him. Illness and indecision delayed Herbert from entering full-time ministry and he was not ordained priest until 1630. Some of Herbert's most poignant and questioning poems of "reversed thunder" were composed during these four "wilderness" years.

Herbert found himself appointed to a small and undistinguished parish church at Bemerton near Salisbury: for just three years he was to exercise his ministry, until his death in 1633. He embraced the life of a parish priest with extraordinary devotion and dedication, and expressed his ideals for pastoral ministry in his work *The Country Parson*. But he faced different struggles during this period. Now he was no longer fighting against his vocation but, dogged with ill-health, found himself questioning his usefulness. Though he valued the presence of Christ in the scriptures and in the sacraments, he wrestled with a sense of spiritual confusion, the dilemma of unanswered prayer, and found himself echoing the sentiments of Jeremiah and the psalmists.[11]

In *Affliction I* he tells the story of his spiritual journey:

> When first thou didst entice to thee my heart,
>   I thought the service brave

---

10. Quoted in Isaac Walton, "The Life of Mr George Herbert" (1670) in Herbert, *English Works*.

11. It is instructive to set *The Country Parson* side by side with Herbert's *Poems*. In *The Country Parson* we see the high ideals of parish ministry: the highest standards of pastoral care visiting, preaching, study and home life. But in the *Poems* we glimpse the reality, the struggle, the heartache, the *angst* of ministry and how Herbert makes sense of this. This section is indebted to Mayes, *Language of the Soul*.

He experienced a time of happiness in God's service:

> What pleasures could I want, whose King I served,
> 	Where joys my fellows were?

But joys passed to sorrows as he encountered both physical and spiritual distress, and having had enough, he explodes with anger to God:

> Well, I will change the service, and go seek
> 	Some other master out.

Herbert experienced prayer as the place of utter transparency before God. In prayer there is no place for false pleasantries, no place for masks, no pretending. In prayer we come before God just as we are, we lower our self-protective barriers, those shields we put up to protect ourselves from others. As a Christian minister, Herbert comes before God with all his woundedness and fragility, all his questions. In his poem *Evensong*, Herbert the parson, at the end of a day of ministry, collapses into his stall in the parish church and takes stock of the successes of the day:

> What have I brought thee home
> For this thy love? have I discharged the debt,
> 	Which this day's favor did beget?
> I ran; but all I brought, was foam.
> 	Thy diet, care, and cost
> Do end in bubbles, balls of wind;
> 	Of wind to thee whom I have crossed,
> But balls of wild-fire to my troubled mind.

Herbert encourages us here to bring to God whatever we face—struggles, burdens or questions—after a day of ministry or giving out. In such metaphors, we can tell God about our struggle for holiness; our perception of failure; a sense of unworthiness; a sense of frustration, in not being effective or successful, not having accomplished what we wanted. We can bring to God a sense of guilt regarding those things "left undone," people unvisited. We can surrender to God our financial worries, health worries, or family concerns. Maybe we will be able to join Herbert as he concludes his *Evensong* prayer:

> My God, thou art all love.
> 	Not one poor minute escapes thy breast,
> 	But brings a favor from above;
> And in this love, more than in bed, I rest.

In the poem *The Collar* Herbert suggests that he is like a restless, rebellious and wayward dog or horse, reluctant to submit himself to the collar or yoke of Christian disciplines ("good cable to enforce and draw"). In it we trace a movement from resistance and protest to ultimate surrender. He begins by thundering a sense of desperation to God:

> I struck the board, and cried, No more;
> > I will abroad.
> What? shall I ever sigh and pine?

After complaining that his only harvest is one of thorns, and his spiritual life is seemingly fruitless, he pauses and catches the echo of God's voice:

> But as I raved and grew more fierce and wild
> > At every word,
> Me thought I heard one calling, *Child*;
> > And I replied, *My Lord*.

The "reversed thunder" of his prayer receives an answering reply. Though his spiritual life might be turbulent, underpinning it all is the fundamental, unchangeable reality: he *is* God's child, and he is held in God's love. In *Longing* he comes to see this clearly as the basic truth of his identity. After laying bare his soul's torments, he confesses that to him God is absent, aloof, faraway, and unresponsive:

> With sick and famished eyes,
> With doubling knees and weary bones,
> > To thee my cries
> > To thee my groans,
> To thee my sighs, my tears ascend:
> > > No end?
> [. . .]   Thou tarriest, while I die,
> And fall to nothing: thou dost reign
> > And rule on high
> > While I remain
> In bitter grief; yet am I styled
> > Thy child.

Here we see a pilgrimage from despair to a new affirmation and sense of identity as God's beloved one. Herbert teaches us about *movement* in prayer: a movement from questions, burdens, struggles to a place of surrender, an end to resisting, as we give in to God. At that point of submission, and at that point alone, we discover God's healing and affirmation. This is explored in the language of such Psalms as 22 and 42. We too can experience transformation, changes in perception, taking place as we pray. Thunder turns to healing, and raving at God turns to listening.

LISTENING

The practice of contemplative listening becomes a "transferable skill" which equips us to relate meaningfully to our contemporary culture. Meaningful interaction with others entails attentive listening to the Other. Listening requires a focused attentiveness to the Other. A double listening is required before any speaking, any evangelism, a listening to God and a discovery of the hopes and hurts in our community—as the report *mission-shaped church* puts it: "listening to the culture . . . and to the inherited tradition of the gospel and the church."[12] We need to attune ourselves both to what God is saying in the historic riches of spirituality, for example, and also to what God is saying in the people of our society today.

First we learn to listen inwardly . . .

- to our own heart, our feelings and responses to God
- to the inner word of God: his whispers and intimations
- to the word of God in Scripture to us

But we also learn to listen to what God is saying to us outwardly

- in the cries of the poor
- in the screams of the oppressed
- in the sobs of the broken-hearted
- in the sighs of our culture
- in the laughter in people's lives

Today we face noise pollution almost everywhere—not only the roar of traffic, and the blast of music but a thunderous cacophony of competing demands and voices clamoring for attention. It is difficult to discern the voice of God, but we need to learn to listen to what "the Spirit is saying to the churches" and to read the "signs of the times." Richard Foster encourages us to: "meditate upon the events of our time and to seek to penetrate their significance. We have a spiritual obligation to penetrate the inner meaning of events and political pressures, not to gain power, but to gain prophetic perspective."[13] As we discipline ourselves to the practice of such discipline, we discover where people are hurting, and where human dignity is being eroded. We echo Isaiah's comment: "The Lord God has opened my

---

12. Archbishops' Council, *mission-shaped church*, 104.
13. Foster, *Celebration of Discipline*, 28.

ear" (Isa 50:5). The Scripture is emphatic: "O that today you would listen to his voice!" (Ps 95:7).

Sometimes God speaks to us through unexpected means. C.S. Lewis put it powerfully:

> God whispers to us in our pleasures, speaks to us in our conscience, but shouts to us in our pains: it is his megaphone to rouse a deaf world . . . No doubt Pain as God's megaphone is a terrible instrument; it may lead to final and unrepented rebellion. But it gives the only opportunity the bad man can have for amendment. It removes the veil; it plants the flag of truth within the fortress of a rebel soul.[14]

Lewis suggests that the experience of pain can shatter the illusion that all is well with us, destroying the false idea that we can get very nicely by without God. Pain shatters the illusion of self-sufficiency, for it causes us to reach out to God either in petition or complaint. In it God thunders to us and wakes us up to the big questions of God and evil, and can draw us into a new surrender to God, the communion for which we were created. Suffering does indeed has a revelatory character, for those with eyes to see it. God speaks most powerfully through the experience of poverty and pain, calling us to simplicity and trust. God is thundering to us in the big issues of our time—are we listening?

QUESTIONS FOR REFLECTION

1. Why is it, do you think, that Christians often prefer talkative prayer rather than listening prayer?
2. Do you find silence inviting or intimidating?
3. "I see the sights that dazzle, the tempting sounds I hear" (from the hymn "O Jesus I have promised"). What sounds tempt you? Which sounds intrigue you? Which disturb you?
4. Where and how do you hear God?
5. How can you retune your spiritual antennae to hear the whisper of God in all things?
6. When was the last time you spoke out courageously in protest at some injustice—and thundered with righteous anger? What did that feel like?

14. Lewis, *Problem of Pain*, 81, 83.

PRAYER EXERCISE

> Morning by morning he wakens—
>   wakens my ear
>   to listen as those who are taught. (Isa 50:4)

Either

Take a prayer walk, with the intentionality of listening acutely and keenly. What sounds do you notice? Attune yourself to the environment, natural or human-made. Practice attentiveness . . . .to the hum of bees, whisper of breeze, the echo of the ricochet of voices, growl of traffic, the padding of your own footprints and the rustle of your clothes . . . . What do you hear of God in all this? Linger and listen. Try to detect how your heart responds. Conclude with:

> O let me hear thee speaking
> in accents clear and still,
> above the storms of passion,
> the murmurs of self-will;
> O speak to reassure me,
> to hasten or control;
> O speak, and make me listen,
> thou guardian of my soul. (John Ernest Bode, 1868).

Or

Imagine you can hear the distant rumbling of thunder. It seems faraway, but it is persistent and does not go away. It assaults our eardrums. It disturbs any sense of complacency we might have. It gnaws at our peacefulness. What might it represent? What cries of the oppressed and downtrodden find an echo in your heart? What voices of unrest boom across today's world, though some would have them stifled? Use your prayer time to discern the rumblings of our time. Like the "Sons of Thunder" stand silently in solidarity with those who suffer, reflecting on your response to their plight.

9

Replenishing the Spirit
Rain, Flood and Drought

O God, you are my God, I seek you,
  my soul thirsts for you;
my flesh faints for you,
  as in a dry and weary land where there is no water.

(Ps 63:1)

I will never forget the sight. It warmed and cheered my heart. Living in Jerusalem, I was returning back to the college one October afternoon when the heavens opened in a terrific downpour drumming into the earth. Of course I groaned and quickened my step. But then I saw it: small children, coming out of school to meet their parents—jumping, splashing about gleefully in the swelling puddles. The joy on their faces and the screams of delight! They were loving it! I thought: I should revise my reactions to rain!

The Holy Land is a thirsty land. The very name for Jerusalem, Zion, derives etymologically from "thirsty place." Jesus refers to the rain as God's gift to all, without distinction: "But I say to you, Love your enemies and pray for those who persecute you, so that you may be children of your Father in heaven; for he makes his sun rise on the evil and on the good, and sends rain on the righteous and on the unrighteous" (Matt 5:44-45). He speaks of the power of the rain when it comes: "The rain fell, the floods came, and the winds blew and beat on that house" (Matt 7:25). As we noted, he refers to the drought at the time of Elijah (Luke 4:25,26) and speaks of compassion in quenching people's thirst: "whoever gives even a cup of cold water to one of these little ones in the name of a disciple—truly I tell you, none of these will lose their reward" (Matt 10:42).

The scriptures celebrate God's promise: "I will give you your rains in their season, and the land shall yield its produce, and the trees of the field shall yield their fruit" (Lev 26:4). The Bible begins and closes with rivers of water, bespeaking creation and new creation: "a stream would rise from the earth and water the whole face of the ground, then the Lord God formed man . . . a river flows out of Eden to water the garden" (Gen 2:6,7,10; compare1:1). In the Apocalypse "The Lamb will guide them to the springs of the water of life" (Rev 7:17). The vision concludes: "Then the angel showed me the river of the water of life, bright as crystal, flowing from the throne of God and of the Lamb through the middle of the street of the city. On either side of the river is the tree of life . . . producing its fruit each month; and the leaves of the tree are for the healing of the nations" (Rev 22:1,2).

In the Holy Land, the terrain is parched for much of the year. East-facing slopes on the Mount of Olives turn from winter green to scorched brown. But at the end of October heavy rains begin to fall for a day or two. These are the "early" or "former" rains which open the agricultural year—soil hardened and cracked by unrelenting sun begins to loosen, and ploughing can begin. The "latter rains" make their advent in March and April, and bring joy and the promise of harvest:

> He will give the rain for your land in its season, the early rain and the later rain, and you will gather in your grain, your wine, and your oil . . . The Lord will open for you his rich storehouse, the heavens, to give the rain of your land in its season and to bless all your undertakings. (Deut 11:14, 28:12)

Jeremiah's ancient words seem so contemporary:

> Judah mourns and her gates languish;
> They lie in gloom on the ground,
> and the cry of Jerusalem goes up.
> Her nobles send their servants for water;
> they come to the cisterns;
> they find no water,
> they return with their vessels empty.
> they are ashamed and dismayed and cover their heads,
> because the ground is cracked.

> Because there has been no rain on the land
> the farmers are dismayed;
> they cover their heads. (Jer 14:2–4)

The Psalmist marvels at the gift of rain on a parched land:

> O God, when you went out before your people,
>   when you marched through the wilderness,
> the earth quaked, the heavens poured down rain
>   at the presence of God, the God of Sinai,
>   at the presence of God, the God of Israel.
> Rain in abundance, O God, you showered abroad;
>   you restored your heritage when it languished. (Ps 68:7–9)

The book of Job reveals a sense of wonder about the gift of rain. It is not to be taken for granted:

> Surely God is great, and we do not know him;
>   the number of his years is unsearchable.
> For he draws up the drops of water;
>   he distils his mist in rain,
> which the skies pour down
>   and drop upon mortals abundantly. (36:26–28)
> God thunders wondrously with his voice;
>   he does great things that we cannot comprehend.
> For to the snow he says, "Fall on the earth";
>   and the shower of rain, his heavy shower of rain,
> serves as a sign on everyone's hand,
>   so that all whom he has made may know it. (37:5–7)
> He loads the thick cloud with moisture;
>   the clouds scatter his lightning.
> They turn round and round by his guidance,
>   to accomplish all that he commands them
>   on the face of the habitable world. (37:11,12)

God asks Job:

> Who has cut a channel for the torrents of rain,
>   and a way for the thunderbolt,
> to bring rain on a land where no one lives,
>   on the desert, which is empty of human life,
> to satisfy the waste and desolate land,
>   and to make the ground put forth grass?
> Has the rain a father,
>   can you lift up your voice to the clouds,
>   so that a flood of waters may cover you?

> Who has the wisdom to number the clouds?
> Or who can tilt the waterskins of the heavens,
>> or who has begotten the drops of dew? (38:25–28,37)

METAPHORS FOR THE DIVINE AND HUMAN

In the Bible the prophets discern in the gift of rain significant aspects of the Divine and human. For Ezekiel they become a sign of God's benediction: "I will make them and the region around my hill a blessing; and I will send down the showers in their season; they shall be showers of blessing" (Ezek 34:26). Isaiah prays:

> Shower, O heavens, from above, and let the skies rain down righteousness; let the earth open, that salvation may spring up, and let it cause righteousness to sprout up also; I the Lord have created it. (Isa 45:8)

Hosea picks up this theme of justice:

> Let us know, let us press on to know the Lord;
>> his appearing is as sure as the dawn;
>
> he will come to us like the showers,
>> like the spring rains that water the earth . . .
>
> Sow for yourselves righteousness; reap steadfast love; break up your fallow ground;
>> for it is time to seek the Lord, that he may come and rain righteousness upon you. (Hos 6:3; 10:12)

But the rain imagery is not only used in reference to the Divine. It can become the language of the soul. Moses offers this prayer about himself and his own ministry:

> May my teaching drop like the rain, my speech condense like the dew; like gentle rain on grass, like showers on new growth. (Deut 32:2)

Later, the prophet Amos will express humanity's vocation: "Let justice roll down like waters, and righteousness like an ever-flowing stream" (5:24).

Isaiah recognizes fierce and unexpected rains as a symbol of enemy attack:

> For you have been a refuge to the poor,
>> a refuge to the needy in their distress,
>> a shelter from the rainstorm and a shade from the heat.

When the blast of the ruthless was like a winter rainstorm.
(Isa 25:4)

Ezekiel sees parched land as a people who have cut themselves off from God's blessing by improper behaviors and injustice: "The word of the Lord came to me: 'Mortal, say to it: You are a land that is not cleansed, not rained upon in the day of indignation'" (Ezek 22:23). And so physical drought reflects the aridity and barrenness of the soul. But if unfruitful deserts become a symbol of humanity's thirst and desperate need for God, the advent of water represents the spiritual renewal and revivification that only God can bring:

> When the poor and needy seek water,
>   and there is none,
>   and their tongue is parched with thirst,
> I the Lord will answer them,
>   I the God of Israel will not forsake them.
> I will open rivers on the bare heights,
>   and fountains in the midst of the valleys;
> I will make the wilderness a pool of water,
>   and the dry land springs of water.
> (Isa 41:17,18; see also 35:1–4,6–7)

In the New Testament, we are called to patience mingled with expectancy:

> Be patient, therefore, beloved, until the coming of the Lord. The farmer waits for the precious crop from the earth, being patient with it until it receives the early and the late rains. You also must be patient. Strengthen your hearts, for the coming of the Lord is near. (Jas 5:6,8)

THE SPIRITUALITY OF WATER

Water has become an evocative symbol of the spiritual life. Baptism reminds us of the need to be flooded, engulfed, drenched and saturated by the waters of the Holy Spirit. Throughout our spiritual life we need the overflowing, inundating Spirit to irrigate the parched earth of our soul. Streams of grace need to percolate the soul. Paul puts it: "God's love has been poured into our hearts through the Holy Spirit that has been given to us" (Rom 5:5).

The Wet Gospel

John's Gospel mentions water in almost every chapter—intermingling its physical necessity with its spiritual significance. Water is needed for the Jewish rites of purification (2:6) and for Christian baptism (1:26, 1:31, 3:23). Jesus says "no one can enter the Kingdom of God unless he is born of water and the Spirit." The Pool of Bethesda to the north of the Temple serves as a healing sanctuary (5:7) while the Pool of Siloam to the south of the Temple becomes a place of restoration (9:7). Water at a wedding becomes the wine-symbol of Christ's passion (2:1–11). The water of the lake becomes a pathway for the Son of God (6:16,19) who can calm its storms. Peter is later to jump into the water of Galilee and swim to Jesus (21:7). Jesus takes water and pours it out to wash the disciples' feet (13:5), a radical symbol of serving others.

The two most significant references to water are found in chapter 4 and chapter 7. To the woman drawing supplies at Jacob's well, Jesus says: "If you knew the gift of God . . . you would have asked him and he would have given you living water . . . those who drink of the water that I will give them will never be thirsty. The water that I will give will become in them a spring of water gushing up to eternal life" (4:14). Jesus promises a Spirit who quenches our deepest thirst, an inner geyser, welling up to eternal life. As the twelfth century Armenian poet puts it:

> O Fountain of life, you asked for water from the woman of Samaria,
> And promised her living water,
> in return for the transitory one.
> Grant to me, O Fountain of Life,
> That holy drink for my soul,
> That flows from the heart in rivers,
> The Spirit from whom grace gushes forth.[1]

And so, chapter seven of John's gospel gives us the dramatic cry of Christ:

> On the last day of the festival, the great day, while Jesus was standing there, he cried out, "Let anyone who is thirsty come to me, and let the one who believes in me drink. As the scripture has said, 'Out of the believer's heart shall flow rivers of living water.'" Now he said this about the Holy Spirit . . . which believers in him were to receive, for as yet there was no Spirit, because Jesus was not yet glorified. (John 7:37–39)

---

1. Kudian, *Nerses Shnorhali*, 45.

The great Jewish festival of Tabernacles, with its vision of the river of God, forms the context of this event. Jesus attends the Temple liturgy where Ezekiel's vision (ch.47) is proclaimed to the pilgrims: a spring of God's generous blessing bursts forth from under the altar of the Temple and spills out to bring renewal to the whole world. The water gets deeper and deeper as Ezekiel follows the line of the river from the holy city out into the desert. At first the prophet can wade in the water, but soon it comes up right to his waist, so he must swim in the river of God's blessing! The Tabernacles festival reached a climax when, on the last day, a solemn ceremony celebrating this vision, carried up to the Temple in a golden vessel waters from the Pool of Siloam, and poured them out as a sign of God's blessing in "the Age to Come." Jesus was watching this ritual when he cried out his urgent, awesome promise. You do not have to wait until the Last Day! With his glorification on the Cross the Spirit will be unleashed as an overflowing stream to renew all of creation. John links the gift of the Spirit to the paschal mystery. At the crucifixion, a fountain of eternal life is opened for humanity: as his side is pierced, blood and water stream out (19:34, compare 1 John 5:6–8). After Jesus is glorified on the Cross, the Spirit can gush.

When Jesus echoes Ezekiel's prophecy and makes his glorious promise, what does he mean? How can the Spirit come to us as a stream of living water and flow in us and out through us? The promise resonates with four aspects of the Spirit's work. First, Jesus alludes to the energy of the Spirit. He is talking of the cascade of the Spirit, the movement of the Spirit, the empowering of the Spirit, his energy within us. Second, he speaks of an inflow and an overflow. The Spirit comes to us and then, bubbling up like a mountain brook, streams out to others. The geography of the Holy Land echoes this message: where there is just an inflow but no out-flow, a receiving but no giving, we get the Dead Sea, in which all life dies. But where there is a healthy inflow and out-flow of the Jordan we get the life-giving Sea of Galilee, teeming with life. Third, Jesus alludes to the cleansing of the Spirit, washing away our mistakes. Part of the work of the Spirit is to apply to us the benefits of the Cross, which makes possible a river of forgiveness and radical new start for these who are penitent. Fourth, Jesus is talking about the renewing and refreshing grace of the Spirit. As sparkling, living water invigorates and enlivens weary bodies, so the Spirit makes us new, replenishing and restoring parched souls: this is the healing grace of the Spirit, echoing Ezekiel's vision of trees with leaves for healing thriving alongside the riverbank (47:12).

The Greening of the Soul

As winter rains replenishing Galilee's reserves become a symbol of God's overflowing grace, so in the history of Christian spirituality the image of water has inspired countless teachers of prayer. Hildegard of Bingen (1098–1179), poet, mystic and musician, celebrates our "greening" or *viriditas*. Today we talk about the "greening of the planet" but nine hundred years ago Hildegard celebrated the presence of the Holy Spirit in the created order through the idea of greening: "greenness is the condition in which earthly beings experience a fulfillment which is both physical and divine; greenness is the blithe overcoming of the dualism between earthly and heavenly." For Hildegard, the wetness or moisture of the planet, revealed in verdant growth, bespeaks the Holy Spirit who "poured out this green freshness of life into the hearts of men and women so that they may bear good fruit."[2] She invites us to see the world differently, overcoming the dichotomy of heaven and earth by glimpsing the heavenly action in the freshness of the planet, which mirrors the human soul. We are being summoned from a pragmatic and self-centered consumer mentality, so deeply entrenched in our culture and mind-set, to seeing creation as not an entity to be manipulated or exploited but a divine presence to be honored.

"I come to my beloved as the dew upon the flower" says Mechthild of Magdeburg (1210–82).[3] In his poem *Thou art indeed just, Lord* Gerard Manley Hopkins prays: "O thou Lord of life, send my roots rain."

St Symeon the New Theologian emphasizes the necessity of personal encounter with the Divine, and tells his own story:

> He led me by the hand as one leads a blind man to the fountain head, that is, to the holy scriptures and to Your divine commandments ... One day when I was hurrying to plunge myself in this daily bath, You met me on the road, You who had already drawn me out of the mire. Then for the first time the pure light of Your divine face shone before my weak eyes ... From that day on, You returned often at the fountain source, You would plunge my head into the water, letting me see the splendor of Your light ... One day when it seemed as though You were plunging me over and over again in the lustral waters, lightning flashes surrounded me. I saw the rays from Your face merge with the waters; washed by these radiant waters, I was carried out of myself ...[4]

---

2. Peter Dronke, quoted in Bowie & Davies, *Hildegard*, 32.
3. Tobin, *Mechthild*, 47.
4. Quoted in Meyendorff, *Palamas*, 49.

For Symeon, the image of the waters becomes a powerful metaphor of the spiritual life, bespeaking the unfathomable resources of the Spirit, and God's generosity in sharing his gifts.

In his hymn *Glorious things of thee are spoken* John Newton puts it:

> See the streams of living waters,
> Springing from eternal love,
> Well supply thy sons and daughters,
> And all fear of want remove;
> Who can faint, while such a river
> Ever flows their thirst t'assuage?
> Grace which, like the Lord, the giver,
> Never fails from age to age.

THE PHYSICALITY OF WATER

Attentiveness to the aridity of soul re-sensitizes us to the physical thirsts on our planet today. In the Holy Land itself, the shortage of water is a critical and divisive issue. Today the Sea of Galilee has a very functional use, as one of Israel's three major water reserves—it is the country's largest reservoir. Via the National Water Carrier its waters are pumped down to the Negev desert for irrigation. Israeli creativity and scientific technology, which invented the drip-system for irrigation, have transformed arid areas into fertile farmlands, watered from the Sea of Galilee. Some years the waters of the lake become alarmingly low, the water line receding at a dramatic rate, the result of poor snows on Hermon, meager rains and overpumping. The water-level has sometimes hovered just above the "black line," the limit of water extraction, and can fall many feet below the desired level. The receding water-level of the Sea of Galilee symbolizes very dramatically a major crisis facing Israel/ Palestine—the desperate clamor for water.

The Struggle for Water

From ancient times people have been striving over water supplies in the Holy Land. The Hebrew scriptures suggest ways in which resources can be shared equably in two stories about disputes over water. In Genesis 21:25–34 we read of Abraham's argument with the Canaanite warlord Abimelech, whose servants seized control of a well of water Abraham was using. Abraham resolves the dispute and avoids further conflict by entering into a covenant with Abimelech, sealing the agreement by his gift to him of seven lambs, which gives the well the name "Beersheba," meaning "the well of the oath."

Some time later Abraham's son Isaac also comes face to face with Abimelech, in a passage which replays close study, Genesis 26:14–33. It vividly illustrates the scramble for water in this thirsty land. Isaac's servants dug a well but Abimelech's herdsmen protest in words that are heard today in this same land: "The water is *ours!*" (Gen 26:20). Isaac refuses to give into fear and intimidation and moves on, sinking another well at a distance (but no doubt using the same underground watertable). Ultimately Isaac enters into a covenant with Abimelech (Gen 26:31), a relationship based on mutual respect and mutual interest—former foes sit and eat and drink (water?) together as they seal a pact and exchange oaths. That same day Isaac's servants come to him rejoicing and say "We have found water!" As Isaac observes (Gen 26:22), there is room in the land for both groups, and there is enough water for all. The way forward from dispute and struggle is by way of covenant and mutual trust. The lessons shout to us from ancient Beersheba!

Dark Waters

But also in the Holy Land flash floods devastate Bedouin settlements in the Judean desert and Jordan valley. Water deposited on the watershed of the Mount of Olives races through the gullies and canyons of the Judean wilderness, rolling boulders at speed, making its descent of 1400 feet to the Jordan valley. This terrifying spectacle brings alive the words of the psalm:

> Save me, O God,
>   for the waters have come up to my neck . . .
> I have come into deep waters,
>   and the flood sweeps over me. (Ps 69:1,2)

But the psalmist is not talking about a literal flood but about the experience of persecution:

> Many are those who would destroy me,
>   my enemies who accuse me falsely. (Ps 69:4)

Such words echo the promise of God in Isaiah:

> When you pass through the waters, I will be with you;
>   and through the rivers, they shall not overwhelm you. (Isa 43:2)

Water is a double-edged symbol: it can denote death and danger, or it can become the sign of salvation. The mystic John of the Cross describes the changes that can take place in prayer through the image of "dark water." From his prison in the Toledo city walls, John could hear few sounds but the rushing waters of the river Tajo below.[5] This became an image to communicate the mysterious way in which God flows into human prayer: "'He made darkness and the dark water his hiding place' (Ps 18:10–11) . . . this darkness . . . and the dark water of his dwelling denote the obscurity of faith in which he is enclosed."[6] The darkness refers to a process of radical dispossession that John sees at the heart of prayer's movement from egocentricity to God-centeredness, a process in which God seeks to reshape us and convert the ego.

This prompts us to become more aware of places on the globe facing the fear and reality of flooding, as icecaps melt and sea levels rise. We think of those low-lying parts of the world that are especially vulnerable to inundation—Bangladesh, Florida, and the Netherlands. We recall that many flood barriers were breached when Hurricane Katrina hit New Orleans in 2005. Fifteen million people were affected in 2017 by floods in China, while heavy monsoon rains regularly flood Pakistan and northern India. The UK expects to see increased winter rainfall. The Guardian reports: "Fears have been raised that the UK could soon see a repeat of the sort of flooding that has hit in recent years after forecasters predicted a one-in-three chance there would be a new record set for monthly rainfall during coming winters."[7]

Across the world 663 million people still do not have access to water; the vast majority of them—over half a billion—live in rural areas.[8] Climate change—felt through extreme weather events such as cyclones, flooding and drought—threatens to worsen their plight. Eighty per cent of the global population already faces threats to its water security. Half a billion people currently live in rural areas without access to safe water. Unpredictable rains pounding the earth destroy fragile infrastructure, while prolonged droughts shrivel up rivers and ponds. Contaminated water and a lack of decent sanitation encourage the spread of diseases such as cholera and malaria. It is the poorest and most marginalized people in society who suffer most. Women and girls walk long distances to collect dirty water, missing out on education, and losing the opportunity to make a living. Was it ever so? As we recall the image of the Woman at the well (John 4), we remember the stories of Genesis 24 where Isaac discovers his beloved Rebecca at the well, drawing water for the family.

---

5. See Matthew, *Impact of God*, 72.
6. Kavanaugh & Rodriguez, *John of the Cross*, 177.
7. Rawlinson, "Winter Rains."
8. Information in this paragraph is drawn from the charity WaterAid.

QUESTIONS FOR REFLECTION

1. What spiritual thirsts exist in your community? What are the tell-tale signs and evidences of such thirsts?
2. How can you respond to these? What spiritual resources do you have to share with others?
3. Which nations of the world face drought? What can you do to help?
4. The homeless and the refugee are exposed to the elements, often drenched to the skin and lashed by the rain. They risk their lives to cross the often-stormy Mediterranean in the quest for freedom. How do you find yourself responding to their plight?
5. In what ways do you experience the Holy Spirit as "living water"? What testimony can you give to others—mirroring that, perhaps, of Symeon the New Theologian?
6. What does the problem of pollution and contamination of water supplies say to you about the spiritual life?

PRAYER EXERCISE

Slowly pour water into a large bowl as a visual focus and reminder of baptism. In his glorious promise of the river of God Jesus suggests three steps his disciples need to take: "Let anyone who is thirsty come to me, and let the one who believes in me drink. As the scripture has said, 'Out of the believer's heart shall flow rivers of living water'" (John 7:37,38). In our prayer we can take these three steps: we thirst, come to Jesus and drink afresh of the Spirit of God.

First acknowledge and recognize your thirst for the Spirit: spend some moments yearning and longing for more of the water of the Spirit in your life. Hold in your imagination the scene of an arid landscape, a thirsty desert representing your soul. Imagine that there are springs and water-courses below you in the ground.

Second, come to Jesus the giver of the Spirit and place yourself in expectant relation to him. Picture Jesus standing by a gushing fountain, springing up from the ground, pointing to it invitingly. Alternatively, open your arms in a gesture of receiving a downpour from heaven, and get ready to be drenched!

Third, we are invited to drink and receive afresh the living water. Imbibe the living waters. Drink in the Spirit. Ask the Holy Spirit to quench and fulfil your deepest thirst.

10

Discovering the Cosmic Christ
Filling all Things

After this I looked, and there in heaven a door stood open! And the first voice, which I had heard speaking to me like a trumpet, said, "Come up here, and I will show you what must take place after this." At once I was in the spirit, and there in heaven stood a throne, with one seated on the throne! And the one seated there looks like jasper and cornelian, and around the throne is a rainbow that looks like an emerald ... Coming from the throne are flashes of lightning, and rumblings and peals of thunder, and in front of the throne burn seven flaming torches, which are the seven spirits of God; and in front of the throne there is something like a sea of glass, like crystal.

(REV 4:4–6)

J.B. Phillips, the New Testament translator, once wrote a book entitled: *Your God is Too Small*. Sometimes the Christian vision can become narrowed. Some focus on understanding first century Palestine in order to set Jesus of Nazareth in context, but he gets, as it were, imprisoned there within a narrow historical frame. Others might focus on Jesus devotionally in such a way that an individualistic "Jesus and me" mentality takes hold. In this chapter we shift from Jesus of Nazareth to the Cosmic Christ, as we retrieve and reconnect to the staggering tradition of the Christ that transcends heaven and earth. We will find this opens before us a breadth of vision that is breathtaking, mind-boggling, stupendous. As a prayer in the Orthodox tradition puts it, said each night at Vespers:

> Heavenly King, Comforter, the Spirit of truth,
> everywhere present and filling all things,
> treasury of good and giver of life,
> come and dwell in us and cleanse us from sin,
> and of your goodness save our souls!

Such a vision of Jesus the Christ and his universal Spirit can revolutionize our approach to mission. Christ is more than just an historical person who walked this earth for 33 years, a great teacher and miracle-worker. A wider vision enables us to glimpse Jesus the Christ as the Alpha and the Omega, filling all things. The Christology or image of Christ that predominates in our thinking and devotion will have a great impact on our view of mission. Is mission a question of seeking more disciples and followers for Jesus of Nazareth? Or might it involve a commitment to the very cosmos we inhabit?

In this chapter, first we look at biblical material, as we ponder the meaning of "The Word made flesh." We engage with the cosmic Christ celebrated in early Christian hymns preserved in the New Testament. Second, we will hear voices from the history of Christian spirituality. Thirdly, we attend to recent writers as we begin to explore the implications of Christology for our contemporary mission.

DISCOVERING THE COSMIC CHRIST IN THE NEW TESTAMENT

Awesome passages in the New Testament expand our consciousness and point us to an astonishing view of Jesus the Christ. Significantly, this cosmic understanding of Christ is found in different authors and across different communities.

The Word was Made Flesh

> In the beginning was the Word, and the Word was with God, and the Word was God. He was in the beginning with God. All

things came into being through him, and without him not one thing came into being. What has come into being in him was life, and the life was the light of all people . . . And the Word became flesh and lived among us, and we have seen his glory. (John 1:1–3,14)

Scholars have long debated the background to John's use of *Logos*, the Word. What does John wish us to keep in mind as he writes of the role of the Word in the creation of the world, the Word that in the fullness of time will be enfleshed and embodied in the person of Jesus? Some point to the Greek background—and this emphasizes the solemn and serious import of the word *Logos*—it represents reason, the rational principle, ensuring order and stability in creation. In Stoic philosophy *Logos* connotes the structuring principle of the universe, a formal abstraction behind created reality. Greek philosophers like Philo thought of the *Logos* as representing divine reason and logic, bringing order into the midst of chaos. It is a staid and stolid approach.

But the Hebrew background to the *Logos* points us to the mysterious and creative Wisdom of God. Ecclesiasticus celebrates God's *Sophia*:

> I came forth from the mouth of the Most High,
>   and covered the earth like a mist.
> I dwelt in the highest heavens,
>   and my throne was in a pillar of cloud.
> Alone I compassed the vault of heaven
>   and traversed the depths of the abyss.
> Over waves of the sea, over all the earth,
>   and over every people and nation I have held sway. (Sirach 24: 3–6)

The *Logos* in the Hebrew scriptures represents God's playmate in the act of creation, God's Wisdom, evoking a joyful, frisky, gamboling, dancing playfulness in the heart of God. Proverbs 8 puts it:

> The Lord created me at the beginning of his work,
>   the first of his acts of long ago . . .
> When he established the heavens, I was there,
>   when he drew a circle on the face of the deep,
> when he made firm the skies above . . .
>   then I was beside him, like a little child
> and I was daily his delight,
>   rejoicing before him always,
>   rejoicing in his inhabited world
>   and delighting in the human race.

This is a dynamic and vibrant view of the *Logos*: the one who is rejoicing in the playfulness of creation, as a little child delights in making new things, crafting and shaping materials. Such an energy, a verve, a daring, a passionate life-force, a sparkling vitality—this is now to be embodied in the person of Jesus.

The Local and the Universal

John's gospel powerfully reveals the paradox we face as we seek to balance in our spiritual practice the local and the universal, the particular and the cosmic. From the outset, it reveals an acute sense of place. The disciples of the Baptist ask him: "Where are you staying?" and he responds invitingly "Come and see." This summons resounds across the gospel. We are invited to share in a journey of discovery. We are invited to accompany Jesus as he traverses the land as a pilgrim and traveler.

We sense that the author of the fourth gospel has a first-hand knowledge of the land, its valleys and plains. John has a sharp eye, a fascination, for the details at such sites, keen attentiveness to physicality and environment. He tells us the well is deep, and it is in a field (4:5,11). There is a lot of water at Aenon near Salim (3:23). Grass covers hills above Sea of Galilee (6:10). Bethany is two miles from Jerusalem (11:2). Lazarus tomb is a cave (11:38). The tomb of Jesus has a low entrance (20:5). We learn details about the Jerusalem Temple: forecourts accommodate a range of animals (2:13–16); the porticoes of Solomon offer shelter in winter (10:22,23); the treasury is a suitable place for teaching (8:20). John takes us to visit Bethesda, describing a pool near the Sheep Gate with five colonnades or porticos (a line of columns supporting a roof-like structure, 5:2). He gives special significance to the Pool of Siloam (9:1–9). The first is north of the Temple, the other south, and both function in relation to the Temple.

While celebrating this alertness to the local and the particular, it is important to recall that all is set within a cosmic perspective: literally so, for when John speaks of the world, he uses the word *cosmos*. For John, there is a paradox in this widest of settings. The cosmos is the object of God's love: "For God so loved the world that he gave his only Son" (3:16). Yet it will reject him and his disciples: "If the world hates you, be aware that it hated me before it hated you. If you belonged to the world, the world would love you as its own. Because you do not belong to the world, but I have chosen you out of the world-therefore the world hates you" (15:18–19). He goes on: "In the world you face persecution. But take courage; I have conquered the world!" (16:33). John's prologue alerts us to this: "The true light, which enlightens everyone, was coming into the world. He was in the world, and the world came into being through him; yet the world did not know him" (1:9,10). Moving between intimacy and ultimacy, John invites us to appreciate the small details of place without losing a sense of the bigger picture—of cosmic dimensions! In the fourth gospel, Jesus is at once the dusty pilgrim and traveler traversing the land, and the very creator Word made flesh!

In today's frantic world we lose a sense of time and place, of sacred space. Globalization and standardization means we may become less attentive to the small picture, the local, the particular and peculiar.[1]

Early Christian Hymns Celebrate Cosmic Christ

Paul's vision is that God will be "all in all" (1 Cor 15:28).

The Letter to the Philippians preserves an early hymn that gives us a glimpse into the increasing expansiveness of vision and understanding experienced by the first Christians:

> Let the same mind be in you that was in Christ Jesus,
> who, though he was in the form of God,
>   did not regard equality with God
>   as something to be exploited,
> but emptied himself,
>   taking the form of a slave,
>   being born in human likeness.
> And being found in human form,
>   he humbled himself
>   and became obedient to the point of death—
>   even death on a cross.
>   Therefore God also highly exalted him
>   and gave him the name
>   that is above every name,
> so that at the name of Jesus
>   every knee should bend,
>   in heaven and on earth and under the earth,
> and every tongue should confess
>   that Jesus Christ is Lord,
>   to the glory of God the Father. (Phil 2: 5–11)

In this great poem the first Christians hold together in a taut tension the historical reality of Jesus of Nazareth and a cosmic view of his divinity. Within a few lines the brutality of the crucifixion is recalled as the center-point of

---

1. For further exploration of this theme see Mayes, *Sensing the Divine*.

history moving to a cosmic worshipping of Christ. The hymn challenges us to hold together the historical and the cosmic in the same breath, and never to see the Cross in terms of a local crucifixion without the wider, mind-boggling perspective and expansiveness of vision.

"Every knee should bend, in heaven and on earth and under the earth": Paul's hymn evokes the Hebrew Bible's three-part world, with the heavens (*shamayim*) above, earth (*eres*) in the middle, and the underworld (*sheol*) below. In the Hebrew scriptures the word *shamayim* represents both the sky/atmosphere, and the dwelling place of God. The *raqia* or firmament—the visible sky—is a solid inverted bowl over the earth, colored blue from the heavenly ocean above it.[2]

Other strands in the New Testament push out the boundaries of our thinking about the person of Christ. Some envisage multiple heavens. The **Letter to the Hebrews** celebrates:

> We have a great high priest who has passed through the heavens, Jesus, the Son of God. (Heb 4:14)
> It was fitting that we should have such a high priest, holy, blameless, undefiled, separated from sinners, and exalted above the heavens. (Heb 7:26)

It sets before us a colossal image in its opening words:

> Long ago God spoke to our ancestors in many and various ways by the prophets, but in these last days he has spoken to us by a Son, whom he appointed heir of all things, through whom he also created the worlds. He is the reflection of God's glory and the exact imprint of God's very being, and he sustains all things by his powerful word. (Heb 1:1–3)

The Letter to the Ephesians speaks of Christ filling multiple heavens in its hymn:

> He has let us know the mystery of his purpose,
> the hidden plan he so kindly made in Christ from the beginning
> to act upon when the times had run their course to the end:
> that he would bring everything together under Christ, as head,
> everything in the heavens and everything upon earth
> (Eph 1: 9–10, JB)

The Letter goes on to offer a prayer that encourages its readers to hold onto and live by such a transformative vision:

---

2. For an account of developing concepts of heaven, see Wright, *Early History of Heaven*. See also Simon, *Ascent to Heaven*.

> May the God of our Lord Jesus Christ, the Father of glory, give you a spirit of wisdom and perception of what is revealed, to bring you to full knowledge of him. May he enlighten the eyes of your mind so that you can see what hope his call holds for you, what rich glories he has promised the saints will inherit and how infinitely great is the power that he has exercised for us believers.

It offers the grounds and basis for such hope:

> This you can tell from the strength of his power at work in Christ, when he used it to raise him from the dead and made him sit at his right hand, in heaven, far above every Sovereignty, Authority, Power, or Domination, or any other name that can be named, not only in this age but also in the age to come. He has put all things under his feet, and made him, as the ruler of everything, the head of the Church; which is his Body, the fullness of him who fills the whole creation. (Eph 1: 17–23)

In cosmic language, the Letter to the Ephesians celebrates Christ filling all things:

> When it says, "He ascended," what does it mean but that he had also descended into the lower parts of the earth? He who descended is the same one who ascended far above all the heavens, so that he might fill all things.

And it goes on to use meteorological metaphors:

> We must no longer be children, tossed to and fro and blown about by every wind of doctrine...
> Be angry but do not sin; do not let the sun go down on your anger (Eph 4: 9–10, 14, 26, JB)

Colossians in its poem opens before our imagination the widest horizons:

> He is the image of the unseen God
> and the first-born of all creation,
> for in him were created
> all things in heaven and on earth:
> everything visible and everything invisible,
> Thrones, Dominations, Sovereignties, Powers –
> all things were created through him and for him.
> Before anything was created, he existed,
> and he holds all things in unity...
> God wanted all perfection

> to be found in him
> and all things to be reconciled through him and for him,
> everything in heaven and everything on earth . . .
> (Col 1:15–17,19–20, JB)

CHRISTIAN THEOLOGIANS AND MYSTICS DISCOVER A COSMIC CHRIST

Through the history of Christian spirituality mystics and teachers have expanded our consciousness of the person of Christ.

Theologians of the Fourth and Fifth Centuries

Seeking to counter a watered-down understanding of Christ as espoused by Arius, outstanding defenders of the faith affirm that Christ is the eternally pre-existent Word of God. They embody their convictions in the Nicene Creed of 325:

> We believe in one Lord, Jesus Christ,
> the only Son of God,
> eternally begotten of the Father,
> God from God, Light from Light,
> true God from true God,
> begotten, not made,
> of one Being with the Father.
> Through him all things were made.

The cosmic Christ reveals the meaning of the universe. Basil of Caesarea declares: "The Word of God . . . pervades the creation." Gregory of Nazianzus says: "This name *Logos* was given to him because he exists in all things that are." Athanasius speaks of "The Logos of God who is over all and who governs all."[3] In his great book *Christus Victor: an historical study of the three main types of the idea of the atonement* Gustaf Aulen celebrates the classic approach to the Cross which is seen as a battlefield where Christ takes on and defeats humanity's greatest foes and negative, destructive forces: the forces of sin and death, indeed everything that threatens, erodes and undermines the integrity of the cosmos and contributes to its degradation.

---

3. Quotes from Pelikan, *Jesus Through the Centuries*, ch. 5.

Hildegard of Bingen

As we move into the medieval period, we encounter again Hildegard of Bingen, poet, mystic and musician. Hildegard hears Christ say to her:

> I, the fiery life of divine wisdom,
> I ignite the beauty of the plains,
> I sparkle the waters,
> I burn in the sun, and the moon, and the stars,
> With wisdom I order all rightly...
> I adorn all the earth.
> I am the breeze that nurtures all things green...
> I am the rain coming from the dew
> That causes the grasses to laugh with the joy of life.[4]

Mechtild of Magdeburg

In her mystical book *The Flowing Light of the Godhead* the Beguine Mechtild combines intensely personal love for Christ with a cosmic view. He is both bridegroom and lover of the soul and also cosmic Lord: here again, intimacy meets ultimacy. Mechtild expresses the paradoxes in Christology: Jesus among us is a pilgrim trudging across the world, a poor worker, but also omnipresent Lord.[5] In prayer she addresses Christ in tender terms:

> You are the feelings of love in my desire.
> You are a sweet cooling for my breast.
> You are a passionate kiss for my mouth.
> You are a blissful joy of my discovery.

But she also delights in the cosmic imagery of the Sun:

> The sparkling sun of the living Godhead
> Shines through the bright water of cheerful humanity...
> You shine into my soul
> Like the sun against gold.

And she hears God responding:

> When I shine, you shall glow.
> When I flow, you shall become wet.[6]

---

4. Uhlein, *Meditations*, 31. Another translation in Fox, *Divine Works*, 8. See also Fox, *Cosmic Christ*, 110.

5. Tobin, *Mechthild*, 286, 284.

6. Tobin, *Mechthild*, 152, 76.

Meister Eckhart

In the 13th century Eckhart writes daringly of giving birth to Christ from a naked immersion in Godhead. The virgin-soul becomes a wife to the Divine and embodies and exudes the very compassion and justice of God, within which she is immersed in this most intimate union with God. He writes:

> From all eternity
> > God lies on a maternity bed
> > Giving birth.
> > The essence of God is birthing...

And he asks:

> What good is it to me
> for the Creator to give birth to his/her son
> if I do not also give birth to him
> in my time
> and my culture?[7]

For Eckhart, God is ever creative, constantly creating. Creation is the outgoing, overflowing ebullience and creativity of God in Christ, to be greeted in the theophanies of the heavens and earth, and in the deep, imaginative powers of the soul: "I have often said, God is creating the whole world now this instant. Everything God made six thousand years ago and more when He made this world, God is creating now all at once."[8] Rejecting pantheism but affirming panentheism, Eckhart encourages us to encounter, greet and serve the Cosmic Christ in every element of creation, at every moment.

RECENT AND CONTEMPORARY PERSPECTIVES

Pierre Teilhard de Chardin (1881–1955), scientist and Jesuit, became captivated by an all-encompassing vision of the cosmic Christ, existing in all things and sustaining all things. As he looked at the material world, his vision, his seeing, was inspired by the text above in Colossians 1:15–17. He came to see Christ present in all things and as the fundamental principle of unity in a fragmenting world. He wrote: "Christ, through his Incarnation, is interior to the world, rooted in the world even in the very heart of the tiniest atom."[9] He was convinced: "Nothing seems to me more vital, from

---

7. Fox, *Meister Eckhart*. 88, 81.
8. Quoted in Woods, *Eckhart's Way*, 95.
9. de Chardin, *Science and Christ*, 36.

the point of view of human energy, than the appearance and, eventually, the systematic cultivation of such a cosmic sense."[10] Inspired by Paul's vision "that God may be all in all" (1Cor 15:28), he believed that the world was in a state of continuous evolution understood as the divinization of the universe. He prayed "that in every creature I may discover and sense you, I beg you: give me faith."[11] But celebrating the divine presence in every creature, and welcoming the Cosmic Christ everywhere does not come without an intense struggle:

> A tremendous spiritual power is slumbering in the depth of our multitude, which will manifest itself only when we have learnt to break down the barriers of our egoisms and, by a fundamental recasting of our outlook, raise ourselves up to the habitual and practical vision of universal realities.
> Jesus, Savior of human activity to which you have given meaning, Savior of human suffering to which you have given living value, be also the Savior of human unity: compel us to discard our pettiness, and to venture forth, resting on you, into the uncharted ocean of charity.[12]

Contemporary spiritual writer Ronald Rolheiser recalls:

> Pierre Teilhard de Chardin was once called to Rome and asked to clarify certain issues in regards to his teachings. At one point, he was asked: "What are you trying to do?" His answer, in effect: "I am trying to write a Christology that is wide enough to incorporate Christ. Christ isn't just an anthropological phenomenon with significance for humanity, but Christ is also a cosmic event with significance for the planet."

This concept challenges the imagination, implying far, far more than we normally dare think. Among other things, it tells us that Christ lies not just at the root of spirituality and morality, but at the base of physics, biology, chemistry, and cosmology as well. This has many implications:

First of all, it means that the spiritual and the material, the moral and the physical, the mystical and the hormonal, and the religious and the pagan do not oppose each other but are part of one thing, one pattern, all infused by one and the same spirit, all drawn to the same end, with the same goodness and meaning. Simply put, the same force is responsible both for the law of gravity and the Sermon on the Mount and both are binding for the

---

10. de Chardin, *Human Energy*, 130–131.
11. de Chardin, *Hymn of the Universe*, 29.
12. de Chardin, *Le Milieu Divin*, 145, 146.

same reason. . . . ultimately part of one and the same thing, the unfolding of creation as made in the image of Christ and as revealing the invisible God.

The fact that Christ is cosmic and that nature is shaped in his likeness means too that God's face is manifest everywhere. If physical creation is patterned on Christ, then we must search for God not just in our scriptures, in our saints, and in our churches . . . if Christ is also the pattern according to which the universe itself is unfolding, then what's good and what's inside of God is also somehow manifest in the raw energy, color, and beauty of the physical . . . Christ is bigger than the historical churches, operates beyond the scope of historical Christianity, and has influences prior and beyond human history itself. It is Christ, visible and invisible—the person, the spirit, the power, and the mystery—who is drawing all things, physical and spiritual, natural and religious, non-Christian and Christian, into one . . .

Teilhard was right. We need a Christology wide enough to incorporate the whole Christ and our imaginations need still to be stretched.[13]

Fresh visions

Prayer invites us to catch a glimpse concludes of the bigger picture, the wider vision, the broader vista, which will give renewed meaning and purpose to our lives. We conclude with a look at a recent writer, who finds in spirituality clues for a greater vision, moving from Jesus of Nazareth to the cosmic Christ. Franciscan **Ilia Delio** concludes her work *Christ in Evolution*: "Life in Christ can never be private or isolated, for Christ is the Word of the Father and the source of the Spirit. Christ is relational by definition and hence the source of community. To live Christ is to live community; to bear Christ in one's life is to become a source of healing love for the sake of community . . ."

Standing as a scientist and Christian within the tradition of de Chardin she concludes: "We must liberate Christ from a Western intellectual form that is logical, abstract, privatized and individualized . . . Christ is the power of God among us and within us, the fullness of the earth and of life in the universe . . . We can look forward toward that time when there will be one cosmic person uniting all persons, one cosmic humanity uniting all humanity, one Christ in whom God will be all in all."[14] How big is your vision? Is your God too small?

---

13. Rolheiser, "Cosmic Christ."
14. Delio, *Christ in Evolution*, 180.

INTO THE FUTURE

Cosmology points to eschatology:

> Look! He is coming with the clouds;
> every eye will see him (Rev 1:7)

The Bible begins and ends with a vision of the world, a cosmology. It moves between two awesome affirmations:

> In the beginning God created the heavens and the earth . . . the Spirit of God was moving over the face of the waters. (Gen 1:1,2)

> Then I saw a new heaven and a new earth; for the first heaven and the first earth had passed away, and the sea was no more . . . (Rev 21:1)

In between, we face the reality of a fractured planet, threatened by degradation, as expressed so powerfully by Paul:

> For the creation waits with eager longing for the revealing of the children of God; for the creation was subjected to futility, not of its own will but by the will of the one who subjected it, in hope that the creation itself will be set free from its bondage to decay and will obtain the freedom of the glory of the children of God. We know that the whole creation has been groaning in labor pains until now; and not only the creation, but we ourselves, who have the first fruits of the Spirit, groan inwardly while we wait for adoption, the redemption of our bodies. (Rom 8:19–23)

Yet while admitting our transience and mortality Paul is able to affirm:

> I consider that the sufferings of this present time are not worth comparing with the glory about to be revealed to us . . .
> I am convinced that neither death, nor life, nor angels, nor rulers, nor things present, nor things to come, nor powers, nor height, nor depth, nor anything else in all creation, will be able to separate us from the love of God in Christ Jesus our Lord. (Rom 8:18,38–39)

With a confidence in the Cosmic Christ we venture towards the city of God:

> The city has no need of sun or moon to shine on it, for the glory of God is its light, and its lamp is the Lamb . . . And there will be no night there. (Rev 21:23,24)

Ultimately, this glorious and fragile planet and its ecology looks forward to a future where God will truly be "all in all":

> The One who was seated on the throne said, "See, I am making all things new ... I am the Alpha and the Omega, the beginning and the end ... " (21:5,6)

QUESTIONS FOR REFLECTION

1. How can we maintain an attentiveness to the local without becoming excessively parochial in outlook? On the other hand, how can we live in solidarity with suffering peoples in different parts of the earth, while remaining rooted in our own locale? What clues do you get from John's gospel?

2. How do you find yourself responding to this expansive view of Christ, celebrating his theophany in atom and galaxy? What implications do you see for our understanding of mission?

3. What do you think are the implications for spiritual practice of an all-encompassing cosmic understanding of Christ? In what ways can a narrower and more historical focus lead to intense action at the local level or limit or impede an appreciation of the vastness of the ecological issues that now face us?

4. How do you find yourself reacting to Rolheiser's take on the Cosmic Christ?

5. What has challenged you most from reading this book? What is God saying to you about your present vocation?

PRAYER EXERCISE

Either

Go outside.

Look down. Reflect: whatever surface you are standing on—grass or concrete or whatever—your feet are planted on Christ. Realize: wherever you are, you are standing on holy ground.

Look up. If at night feel the blackness and expansiveness of the starlit universe. Reflect: you are gazing on Christ. By day, lift up your eyes to the skies and notice what the heavens are saying to you. Reflect: you are listening to Christ.

Look within. Listen to your heartbeat and pattern of your breathing. Reflect: it is the breath of Christ.

After some moments, use these words attributed to St Patrick:

> Christ be with me,
> Christ within me,
> Christ behind me,
> Christ before me,
> Christ beside me,
> Christ to win me,
> Christ to comfort
> and restore me.
> Christ beneath me,
> Christ above me,
> Christ in quiet,
> Christ in danger,
> Christ in hearts of
> all that love me,
> Christ in mouth of
> friend and stranger.

Conclude by saying very slowly these words from the Easter Vigil:

> Christ yesterday and today
> The Beginning and the End
> The Alpha and the Omega.
> All time belongs to him and all the ages.
> To him be glory and power through every age and for ever.
> Amen.

Or

Remaining inside, arrange on a central table items from the natural world: flowers, leaves, rocks, stones, wood, seeds. If in a group, invite each participant to come forward and take an item to hold gently in their hands. Invite them to consider its fragility or strength, its enduring characteristics or its vulnerability. End by inviting each one to offer a short prayer reflecting our responsibility for care and protection of the planet.

Bibliography

Allen, Rosamund, trans. *Richard Rolle: English Writings*. New York: Classics of Western Spirituality, Paulist, 1988.
Archbishops' Council, *Mission-Shaped Church*. London: Church House, 2004.
Armstrong, Regis J., Hellman, J.A. Wayne and Short, William J., eds. *Francis of Assisi Early Documents: Vol. 1, The Saint*. New York: New City Press, 2000.
Aulen, Gustaf. *Christus Victor: an Historical Study of the Three Main Types of the Idea of the Atonement*. London: SPCK, 1970.
Avis, Paul. *God and the Creative Imagination: Metaphor, Symbol and Myth in Religion and Theology*. London: Routledge, 1999.
Bammel, E. & Moule. C.F.D., eds. *Jesus and the Politics of his Day*. Cambridge University Press, 1984.
Barry, William A. & Connolly, William J. *The Practice of Spiritual Direction*. New York: Seabury, 1982.
Bowie, Fiona & Davies, Oliver, eds. *Hildegard of Bingen: An Anthology*. London: SPCK, 1990.
Brock, Sebastian, trans. *St. Ephrem the Syrian: Hymns on Paradise*, Crestwood, New York: St. Vladimir's Seminary, 1990.
———.*The Luminous Eye: The Spiritual World Vision of St Ephrem the Syrian*. Kalamazoo, Michigan: Cistercian, 1992.
Brown, Robert M. *Spirituality and Liberation*. London: Spire, 1988.
Campbell, Joseph. *Thou art That: Transforming Religious Metaphor*. Novato, CA: New World Library, 2013.
de Chardin, Teilhard. *Human Energy*. New York: Harcourt Brace Jovanovich, 1969.
———. *Science and Christ*. San Francisco: Harper & Row, 1968.
———. *Le Milieu Divin: an Essay on the Interior Life*. London: Fontana, 1970.
———. *Hymn of the Universe*. London: Collins, 1970.
Colledge, Edmund & Walsh, James, trans. *Julian of Norwich—Showings*. New York: Paulist, 1978.
Cousins, Ewert, trans. *Bonaventure: The Soul's Journey into God*. New York: Paulist, 1978.
Crosby, Michael H. *House of Disciples: Church, Economics and Justice in Matthew*. Maryknoll, New York: Orbis, 1988.

Dando, William A. "Clouds and the Promises of God." In *Geography of the Holy Land: Perspectives,* edited by William A. and Caroline Z. Dando, and Jonathan J. Lu, 61–66. Taiwan: Kaohsiung Holy Light Theological Seminary, 2005.

Delio, Ilia. *Christ in Evolution*. Orbis, 2008.

Dunn, James D.G. *Jesus Remembered*. Michigan: Eerdmans, 2003.

Erdman, David. ed. *The Complete Poetry and Prose of William Blake*. University of California Press, 2008.

Fagan, Brian. *The Little Ice Age: How Climate Made History 1300–1850*. New York: Basic, 2000.

Foster, Richard. *Celebration of Discipline*. London: Hodder and Stoughton, 1980.

Fox, Matthew. *Meditations with Meister Eckhart*. Rochester, Vermont: Bear & Co., 1983.

———. *Hildegard of Bingen's Book of Divine Works*. Rochester, Vermont: Bear & Co., 1987.

———. *The Coming of the Cosmic Christ*. San Francisco: HarperOne, 1990.

———. *Illuminations of Hildegard of Bingen*. Rochester, Vermont: Bear & Co., 2002.

Francis, Pope. *Laudato Si: On Care of our Common Home*. New York: Paulist, 2015.

French, R.M. trans. *The Way of a Pilgrim*. London: SPCK, 2012.

Frye, Northrop. *The Double Vision: Language and Meaning in Religion*. Toronto: University of Toronto Press, 1991.

Gillet, Lev. *The Jesus Prayer*. Oxford: Mowbrays, 1987.

Guterres, Antonio. "UN Secretary General's Cop26 statement 13 November 2021." www.thenationalnews.com/world/cop-26/2021/11/13/full-text-of-un-secretary-generals-cop26-statement/

Harris, A. *Weatherland: Writers & Artists under English Skies*. London: Thames & Hudson, 2015.

Hedahl, Susan. *Listening Ministry*. Minneapolis: Fortress, 2001.

Herbert, George. *The Complete English Works*. London: David Campbell Publishers, Everyman's Library, 1995.

Hiebert, Theodore. *The Yahwist's Landscape: Nature and Religion in Early Israel*. Minneapolis: Fortress, 1996.

Hillel, David. *The Natural History of the Bible*. New York: Columbia University Press, 2006.

Hutchinson, Francis E., ed. *The Works of George Herbert*. Oxford: Clarendon, 1970.

Justes, Emma A. *Hearing beyond words: how to become a listening pastor*. Nashville: Abingdon, 2010.

Lakoff, George & Johnson, Mark. *Metaphors We Live By*. Chicago: University of Chicago Press,1980.

Lawrence, D.H. *Complete Poems*. London: Penguin, 1994.

Lewis, C.S. *The Problem of Pain*. London: Fontana, 1976.

Kavanaugh, Kieran & Rodriguez, Otilio, trans. *The Collected Works of St John of the Cross*. Washington DC: ICS, 1991.

Kelsey, David H. "Reflections on a Discussion of Theological Education as Character Formation," *Theological Education* 25:1 (1988), 64.

Klocek, D. *Climate: Soul of the Earth*. Great Barrington, MA: Lindisfarne Books, 2011.

Kudian, Muscha, trans. *Nerses Shnorhali: Jesus, the Son*. London: Mashtots, 1986.

Long, Ann. *Listening*. London: Darton, Longman and Todd, 1990.

Longfellow, Henry Wadsworth. *The Poetical Works*. Oxford: Oxford University Press, 1908.

Lonsdale, David "The Cloud of Unknowing" in Lavinia Byrne, ed. *Traditions of Spiritual Guidance*. Minnesota: Liturgical, 1990.
Lossky, Vladimir. *The Mystical Theology of the Eastern Church*. London: James Clarke,1957.
Louth, Andrew. *The Origins of the Christian Mystical Tradition*. Oxford: Oxford University Press, 2007.
Luibheid, Colm & Russell, Norman, trans. *John Climacus: The Ladder of Divine Ascent*. New York: Paulist, 1982.
Malherbe, Abraham J. & Ferguson, Everett, trans. *Gregory of Nyssa: The Life of Moses*. New York: Paulist, 1978.
Maloney, George A., trans. *The Mystic of Fire and Light: St. Symeon the New Theologian*. Denville, New Jersey: Dimension, 1975.
———. *Intoxicated with God: The Fifty Spiritual Homilies of Macarius*. Denville, New Jersey: Dimension, 1978).
Matthew, Ian. *The Impact of God: Soundings from St John of the Cross*. London: Hodder & Stoughton, 1995.
Mayes, Andrew D. *Learning the Language of the Soul*. Collegeville, MN: Liturgical, 2016.
———. *Sensing the Divine: John's Word Made Flesh*. Abingdon: BRF 2018.
McVey, Kathleen E. *Ephrem the Syrian: Hymns*. New York: Paulist, 1989.
Merton, Thomas. *When the Trees say Nothing: Writings on Nature*. Notre Dame, IN: Sorin, 2003.
Meyendorff, John. *St Gregory Palamas and Orthodox Spirituality*. New York: St Vladimir's Seminary, 1974.
Moltmann, Jurgen. *The Spirit of Life: An Universal Affirmation*. Minneapolis: Fortress Press, 2001.
Monbourquette, John. *How to Befriend Your Shadow: Welcoming Your Unloved Side*. London: Darton, Longman & Todd, London, 2001.
Mottola, Anthony, trans., *The Spiritual Exercises of St Ignatius*. New York: Image/Doubleday, 1964.
Mulholland, M. Robert. *Invitation to a Journey: A Road Map for Spiritual Formation*. Intervarsity, 1993.
Oliver, John W. *Giver of Life: The Holy Spirit in Orthodox Tradition*. Brewster, MA: Paraclete.
Pelikan, Jaroslav. *Jesus Through the Centuries*. Yale University Press, 1985.
Pennington, Jonathan T. *Heaven and Earth in the Gospel of Matthew*. Leiden: Brill, 2007.
Rawlinson, Kevin. "UK should increasingly expect record winter rains, says Met Office." www.theguardian.com/uk-news/2017/jul/24/uk-winter-rain-met-office-super computer-record-rainfall-flooding
Reveney, Denis. *Language, Self and Love: Hermeneutics in the Writings of Richard Rolle and the Commentaries on the Song of Songs*. Cardiff: University of Wales Press, 2001.
Robinson, Edward. *Biblical Researches in Palestine, Mount Sinai and Arabia Petraea*. Boston: Crocker & Brewster, 1841.
Rolheiser, Ronald. "The Cosmic Christ." www.ronrolheiser.com
Rowlatt, Justin. "COP26: World at one minute to midnight over climate change—Boris Johnson." www.bbc.co.uk/news/uk-59114871
Sheldrake, Phillip. *Befriending Our Desires*. London: Darton, Longman & Todd, 1994.

Sherley-Price, L., trans. *Walter Hilton: The Ladder of Perfection*. Harmondsworth: Penguin, 1957.
Symeon the New Theologian. *Divine Eros: Hymns of St Symeon the New Theologian*. Translated by Daniel K. Griggs. New York: St Vladimir's Seminary Press, 2011.
Simkins, Ronald A. *Creator and Creation: Nature in the Worldview of Ancient Israel*. Peabody, Mass: Hendrickson, 1994.
Simon, Ulrich. *The Ascent to Heaven*. London: Barrie and Rockliffe, 1961.
Smith, George Adam. *The Historical Geography of the Holy Land*. London: Hodder and Stoughton, 1935.
Soelle, Dorothy. *The Inward Road and the Way Back* . London: Darton, Longman & Todd, 1978.
Soskice, Janet M. *Metaphor and Religious Language*. Oxford: Clarendon, 1985.
Steere, Douglas. *On Listening to Another*. San Francisco: Harper, 1955.
Stoutzenberger, Joseph M. & Bohrer, John D. *Praying with Francis of Assisi*. Winona, Minnesota: Saint Mary's, 1989.
Telegraph, "Why Brits Always Talk About the Weather." (20 Feb 2017). www.telegraph.co.uk/news/picturegalleries/howaboutthat/6214281/Why-Brits-always-talk-about-the-weather.html
Thomas, R.S. *Laboratories of the Spirit*. London: Macmillan, 1975.
Thornton, Martin. *English Spirituality: An Outline of Ascetical Theology According to the English Pastoral Tradition*. Cambridge, Massachusetts: Cowley,1986.
Tobin, Frank, trans. *Mechthild of Magdeburg: The Flowing Light of the Godhead*. New York: Paulist, 1998.
Traherne, Thomas. "The Kingdom of God" in Ross, Jan, ed. *The Works of Thomas Traherne*. Cambridge: D. S. Brewer, 2014.
Tuchman, Barbara W. *A Distant Mirror: The Calamitous 14th Century*. Harmondsworth: Penguin, 1978.
Uhlein, Gabriele. *Meditations with Hildegard of Bingen*. Sante Fe NM: Bear & Co, 1983.
Underhill, Evelyn. *Immanence: A Book of Verses*. London: J.M. Dent & Sons, 1915.
Vaughan-Lee, Llewellyn, ed. *Spiritual Ecology: The Cry of the Earth*. Point Reyes Station, California: Golden Sufi Center, 2016.
de Waal, Esther. *A Seven Day Journey with Thomas Merton*. Guildford: Eagle, 2000.
Walker, Peter. *Jesus and his World*. Oxford: Lion, 2003.
Wallis, Jim. *The Soul of Politics*. London: Fount, 1994.
Walsh, John, trans. *The Cloud of Unknowing*. New York: Classics of Western Spirituality, Paulist, 1981.
Ward, Benedicta, trans. *The Sayings of the Desert Fathers: the Alphabetical Collection*. Kalamazoo: Cistercian, 1975.
Westcott, Brooke F. *The Gospel according to St John*. London: John Murray, 1902.
van de Weyer, Robert, trans. *Celtic Fire: An Anthology of Celtic Christian Literature*. London: Darton, Longman and Todd, 1990.
White, Dominic. *The Lost Knowledge of Christ: Contemporary Spiritualities, Christian Cosmology, and the Arts*. Minnesota: Liturgical, 2015.
Wiggins, Steve A. *Weathering the Psalms*. Eugene, Oregon: Cascade, 2014.
Williams, Rowan. *The Edge of Words: God and the Habits of Language*. London: Bloomsbury, 2014.
Wolters, Clifton, trans. *Richard Rolle: The Fire of Love*. Harmondsworth: Penguin, 1972.
———. *The Cloud of Unknowing*. Harmondsworth: Penguin,1976.

Woods, Richard. *Eckhart's Way.* London: Darton, Longman & Todd, 1986.
Wren, Brian. *What Language Shall I Borrow? God –Talk in Worship.* London: SCM, 1989.
Wright, J. Eward. *The Early History of Heaven.* Oxford: Oxford University Press, 2000.

www.ingramcontent.com/pod-product-compliance
Lightning Source LLC
Chambersburg PA
CBHW050825160426
43192CB00010B/1896